Comprehensive

PRACTICAL CHEMISTRY

FOR

CLASS XII

Strictly according to new curriculum prescribed by
Central Board of Secondary Education (CBSE)
and
State Boards of Chhattisgarh, Haryana, Bihar, Jharkhand, Kerala, Mizoram, Meghalaya, Uttarakhand and other States following NCERT Curriculum

By

Dr. N.K. VERMA
Formerly, Associate Professor
Chemistry Department
D.A.V. College
Chandigarh

Dr. B.K. VERMANI
Formerly, Associate Professor
Chemistry Department
D.A.V. College
Chandigarh

Dr. NEERA VERMA
Formerly, Associate Professor
Chemistry Department
M.C.M. D.A.V. College
Chandigarh

K.K. REHANI
Formerly, Lecturer of Chemistry
S.G.G.S. College
Chandigarh

LAXMI PUBLICATIONS (P) LTD
(An ISO 9001:2015 Company)
BENGALURU • CHENNAI • GUWAHATI • HYDERABAD
KOCHI • KOLKATA • LUCKNOW • RANCHI
NEW DELHI

Printed and bound in India
New Edition
ISBN : 978-81-318-0371-4

Branches

✆	Bengaluru	080-26 75 69 30
✆	Chennai	044-24 34 47 26
✆	Guwahati	0361-254 36 69
✆	Hyderabad	040-27 55 53 83
✆	Kochi	0484-405 13 03
✆	Kolkata	033-40 04 77 79
✆	Lucknow	0522-220 99 16
✆	Ranchi	0651-224 24 64

PUBLISHED IN INDIA BY

Laxmi Publications (P) Ltd.
(An ISO 9001:2015 Company)
113, GOLDEN HOUSE, GURUDWARA ROAD, DARYAGANJ,
NEW DELHI - 110002, INDIA
Telephone : 91-11-4353 2500, 4353 2501
www.laxmipublications.com info@laxmipublications.com

C—00000/025/03
Printed at: Ajit Printing Press, Delhi

Comprehensive

PRACTICAL CHEMISTRY

FOR

CLASS XII

WIN A CASH AWARD OF ₹ 500.00

It has come to our notice that some booksellers are fraudulently **selling fake/duplicate copies** of some of our fast selling titles.

In our sincere efforts to provide you with our genuine books and to protect you against these counterfeit books, Laxmi Publications (LP) has put a Hologram on the cover of **some of its fast selling** titles. The Hologram displays a unique **3D Litho, multi-level, multi-colour** effect from different angles. It has the following **eleven levels of flat graphics** merged together. The **3D Depth** artwork seems to be 'under' or 'behind' the Hologram and gives the illusion of depth unlike the fake Hologram on the fake/duplicate books.

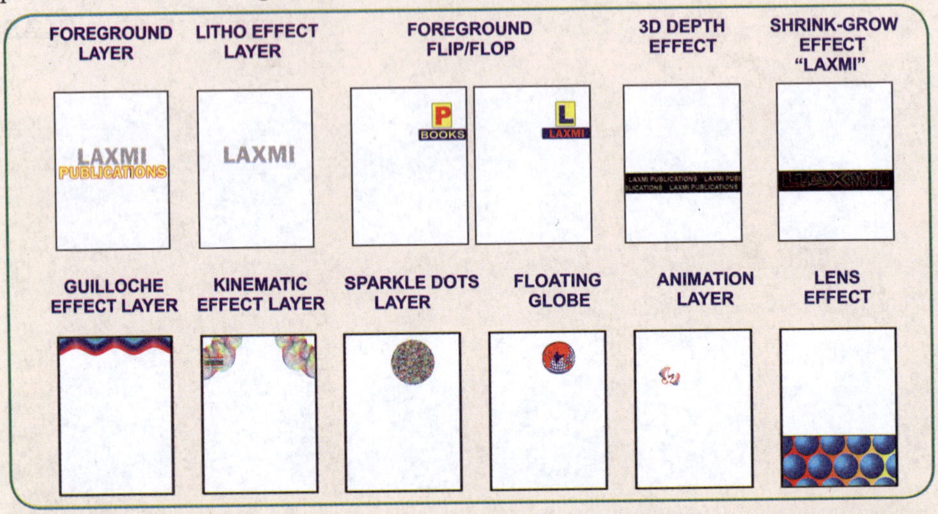

Presently, only some titles have got the Holograms. In this case, ***Comprehensive* Practical Chemistry** for **Class XII** has got the Hologram.

If you or any of your friends finds anywhere in India/abroad any book of this Edition without the LP Hologram, he/she is requested to write to us at **M/s LAXMI PUBLICATIONS PVT. LTD., 113, Golden House, Gurudwara Road, Daryaganj, New Delhi-110002**, giving the name and address of the bookseller from where he/she purchased this book, together with the photocopy of the cover and the 4th page on which the name of the printer is printed, he/she will be sent a **cash award of ₹ 500.00**.

How to decide if the book is genuine or fake?

1. The above information may or may not be printed.
2. The counterfeit edition of the book may have no LP Hologram or if it has, it will be without the illusionary depth as described above.

What is the harm in purchasing duplicate books?

— Poor quality of paper and printing which affects your eyes.

— No royalty to authors who are scholars and have put their hard labour in writing the book, thus depriving them of their intellectual rights.

Warning : Selling or buying pirated books is an offence. Legal action shall be taken against the bookseller(s) and student(s) or whoever found guilty of such an offence in any way.

CONTENTS

PREFACE

We are pleased to present the revised edition of the *'Comprehensive* **Practical Chemistry'** for **Class XII** students. The book has been written strictly according to the new syllabus prescribed by Central Board of Secondary Education. We hope that the book will be quite helpful to the students in acquiring the skills of various laboratory techniques. Some of the outstanding features of the book are:

- Simple language and lucid style.
- Wherever required, a large number of illustrations have been given to clarify the use of various apparatuses used in laboratory.
- The theoretical aspects of each experiment have been discussed briefly along with the experiment.
- In volumetric analysis calculations based on normality as well as on molarity have been given.
- In qualitative inorganic analysis, the various tests have been given in a systematic way in tabular form.
- In order to guide the students about recording the experiment in the notebook, a specimen record of analysis of a salt has been given in the chapter on qualitative analysis.
- A large number of solved viva questions have been included in each chapter.

We sincerely hope that the book will be appreciated by our learned colleagues and students. We shall be glad to receive constructive suggestions for the further improvement of the book.

—AUTHORS

SYLLABUS

CLASS XII (PRACTICALS)

Evaluation Scheme for Examination	Marks
Volumetric Analysis	08
Salt Analysis	08
Content Based Experiment	06
Project Work	04
Class Record and Viva	04
Total	**30**

Practicals Syllabus — Total Periods 60

Micro-chemical methods are available for several of the practical experiments. Wherever possible, such techniques should be used.

A. Surface Chemistry

(*a*) Preparation of one lyophilic and one lyophobic sol.

Lyophilic sol—starch, egg albumin and gum.

Lyophobic sol—aluminium hydroxide, ferric hydroxide, arsenious sulphide.

(*b*) Dialysis of sol—prepared in (*a*) above.

(*c*) Study of the role of emulsifying agents in stabilizing the emulsions of different oils.

B. Chemical Kinetics

(*a*) Effect of concentration and temperature on the rate of reaction between sodium thiosulphate and hydrochloric acid.

(*b*) Study of reaction rates of any one of the following :

(*i*) Reaction of iodide ion with hydrogen peroxide at room temperature using different concentration of iodide ions.

(*ii*) Reaction between potassium iodate (KIO_3) and sodium sulphite : (Na_2SO_3) using starch solution as indicator (clock reaction).

C. Thermochemistry

Any one of the following experiments

(*i*) Enthalpy of dissolution of copper sulphate or potassium nitrate.

(*ii*) Enthalpy of neutralization of strong acid (HCl) and strong base (NaOH).

(*iii*) Determination of enthalpy change during interaction (Hydrogen bond formation) between acetone and chloroform.

D. Electrochemistry

Variation of cell potential in Zn | Zn^{2+} || Cu^{2+} | Cu with change in concentration of electrolytes ($CuSO_4$ or $ZnSO_4$) at room temperature.

E. Chromatography

(i) Separation of pigments from extracts of leaves and flowers by paper chromatography and determination of R_f values.

(ii) Separation of constituents present in an inorganic mixture containing two cations only (constituents having large difference in R_f values to be provided).

F. Preparation of Inorganic Compounds

(i) Preparation of double salt of ferrous ammonium sulphate or potash alum.

(ii) Preparation of potassium ferric oxalate.

G. Preparation of Organic Compounds

Preparation of any one of the following compounds

(i) Acetanilide

(ii) Di-benzal acetone

(iii) *p*-Nitroacetanilide

(iv) Aniline yellow or 2-Naphthol aniline dye.

H. Tests for the functional groups present in organic compounds :

Unsaturation, alcoholic, phenolic, aldehydic, ketonic, carboxylic and amino (primary) groups.

I. Characteristic tests of carbohydrates, fats, and proteins in pure samples and their detection in given food stuffs.

J. Determination of concentration/molarity of $KMnO_4$ solution by titrating it against a standard solution of :

(i) Oxalic acid

(ii) Ferrous ammonium sulphate.

(Students will be required to prepare standard solutions by weighing themselves).

K. Qualitative Analysis

Determination of one cation and one anion in a given salt.

Cations: Pb^{2+}, Cu^{2+}, As^{3+}, Al^{3+}, Fe^{3+}, Mn^{2+}, Zn^{2+}, Co^{2+}, Ni^{2+}, Ca^{2+}, Sr^{2+}, Ba^{2+}, Mg^{2+}, NH_4^+

Anions: CO_3^{2-}, S^{2-}, SO_3^{2-}, SO_4^{2-}, NO_2^-, Cl^-, Br^-, I^-, PO_4^{3-}, $C_2O_4^{2-}$, CH_3COO^-, NO_3^-

(***Note:*** Insoluble salts excluded)

PROJECT

Scientific investigations involving laboratory testing and collecting information from other sources.

A Few Suggested Projects

- Study of the presence of oxalate ions in guava fruit at different stages of ripening.
- Study of quantity of casein present in different samples of milk.
- Preparation of soyabean milk and its comparison with the natural milk with respect to curd formation, effect of temperature, etc.
- Study of the effect of potassium bisulphate as food preservative under various conditions (temperature, concentration, time, etc.)
- Study of digestion of starch by salivary amylase and effect of pH and temperature on it.
- Comparative study of the rate of fermentation of following materials : wheat flour, gram flour, potato juice, carrot juice, etc.
- Extraction of essential oils present in Saunf (aniseed), Ajwain (carum), Illaichi (cardamom).
- Study of common food adulterants in fat, oil, butter, sugar, turmeric powder, chilli powder and pepper.

Note: Any other investigatory project, which involves about 10 periods of work, can be chosen with the approval of the teacher.

Practical Examination for Visually Impaired Students

Evaluation Scheme

Time Allowed: Two hours **Max. Marks: 30**

Identification/Familiarity with the apparatus	5 marks
Written test (based on given/prescribed practicals)	10 marks
Practical Record	5 marks
Viva	10 marks
Total	**30 marks**

General Guidelines

- The practical examination will be of two hour duration.
- A separate list of ten experiments is included here.
- The written examination in practicals for these students will be conducted at the time of practical examination of all other students.
- The written test will be of 30 minutes duration.
- The question paper given to the students should be legibly typed. It should contain a total of 15 practical skill based very short answer type questions. A student would be required to answer any 10 questions.
- A writer may be allowed to such students as per CBSE examination rules.
- All questions included in the question papers should be related to the listed practicals. Every question should require about two minutes to be answered.
- These students are also required to maintain a practical file. A student is expected to record at least five of the listed experiments as per the specific instructions for each subject. These practicals should be duly checked and signed by the internal examiner.
- The format of writing any experiment in the practical file should include aim, apparatus required, simple theory, procedure, related practical skills, precautions etc.
- Questions may be generated jointly by the external/internal examiners and used for assessment.
- The viva questions may include questions based on basic theory/principle/concept, apparatus/materials/chemicals required, procedure, precautions, sources of error etc.

A. Items for Identification/Familiarity of the apparatus for assessment in practicals (All experiments)

Beaker, glass rod, tripod stand, wire gauze, Bunsen burner, Whatman filter paper, gas jar, capillary tube, Pestle and mortar, Test tubes, tongs, test tube holder, test tube stand, burette, Pipette, conical flask, standard flask, clamp stand, burner, funnel, filter paper.

Hands-on Assessment

- Identification/familiarity with the apparatus
- Odour detection in qualitative analysis

B. List of Practicals

The experiments have been divided into two sections: Section A and Section B. The experiments mentioned in Section B are mandatory.

SECTION—A

A. Surface Chemisty

(1) Preparation of one lyophilic sol

Lyophilic sol—starch, egg albumin and gum

(2) Preparation of one lyophobic sol

Lyophobic sol—Ferric hydroxide

B. Chromatography

(1) Separation of pigments from extracts of leaves and flowers by paper chromatography and determination of R_f values (distance value may be provided).

C. Tests for the Functional Groups Present in Organic Compounds

(1) Alcoholic and Carboxylic groups.

(2) Aldehydic and Ketonic

D. Characteristic Tests of Carbohydrates and Proteins in the given Food Stuffs

E. Preparation of Inorganic Compounds—Potash Alum

SECTION—B (Mandatory)

F. Quantitative Analysis

(1) (*a*) Preparation of the standard solution of Oxalic acid of a given volume.

(*b*) Determination of molarity of $KMnO_4$ solution by titrating it against a standard solution of Oxalic acid.

(2) The above exercise [F 1 (*a*) and (*b*)] to be conducted using Ferrous ammonium sulphate (Mohr's salt).

G. Qualitative Analysis

(1) Determination of one cation and one anion in a given salt.

Cation – NH_4^+

Anions – CO_3^{2-}, S^{2-}, SO_3^{2-}, Cl^-, CH_3COO^-

(*Note:* Insoluble salts excluded)

Note: The above practicals may be carried out in an experiential manner rather than recording observations.

CHAPTER 1

Introduction to Basic Laboratory Equipment

1.01. CHEMISTRY LABORATORY

A chemistry laboratory is a workshop for chemists. Here students learn the techniques of the preparation, identification and estimation of chemical substances. Before starting experiment, a student must know from where to get the apparatus required for the given experiment and the placement of the chemicals to be used. A student must know the proper use of each equipment and the precautions to be observed while working in the laboratory. A chemistry laboratory is provided with the following fittings with which the student must become familiar.

1. Demonstration Table

Before starting experiment, the teacher gives instructions and demonstrates the concerned experiment on demonstration table. In chemistry laboratory, no seats are made available to the students, so students stand around demonstration table and note the instructions from teacher.

2. Students' Working Table

A number of wooden or concrete tables are provided for working. Generally, four students (two on each side) work on one table. Each seat is provided with:

(*a*) **Reagent shelves.** Reagents or chemicals to be used are placed on the reagent shelf. These are the reagents which are commonly used. For example, all dilute and concentrated acids such as H_2SO_4, HCl, HNO_3, etc. and bases like NaOH, NH_4OH, etc.

(*b*) **Sinks and water taps.** A sink and a water tap is fitted between every two reagent shelves. On either side of the sink, usually two taps are fitted for supply of water.

(*c*) **Gas taps.** These taps are fitted on the seats for supply of petrol gas to the burners. Sometimes kerosene is used for producing gas in place of petrol.

3. Side Shelves

Mostly there are two big shelves fitted on the walls of the laboratory. Reagents and chemicals, which are less frequently used, are placed in these shelves. Sometimes solid chemicals are placed in a separate shelf.

4. Fume Cup-board

There is at least one fume cup-board in the corner of the laboratory. All experiments giving out poisonous gases or vapours are performed in this cup-board.

5. Balance Room

It is a small room attached to each laboratory. Here, a number of balances are kept for weighing the substances.

6. Exhaust Fans

Two exhaust fans are provided at the two corners of the laboratory for the removal of the poisonous gases and vapours from the laboratory.

1.02. COMMON LABORATORY APPARATUS

The apparatus which is commonly used by a student is described below:

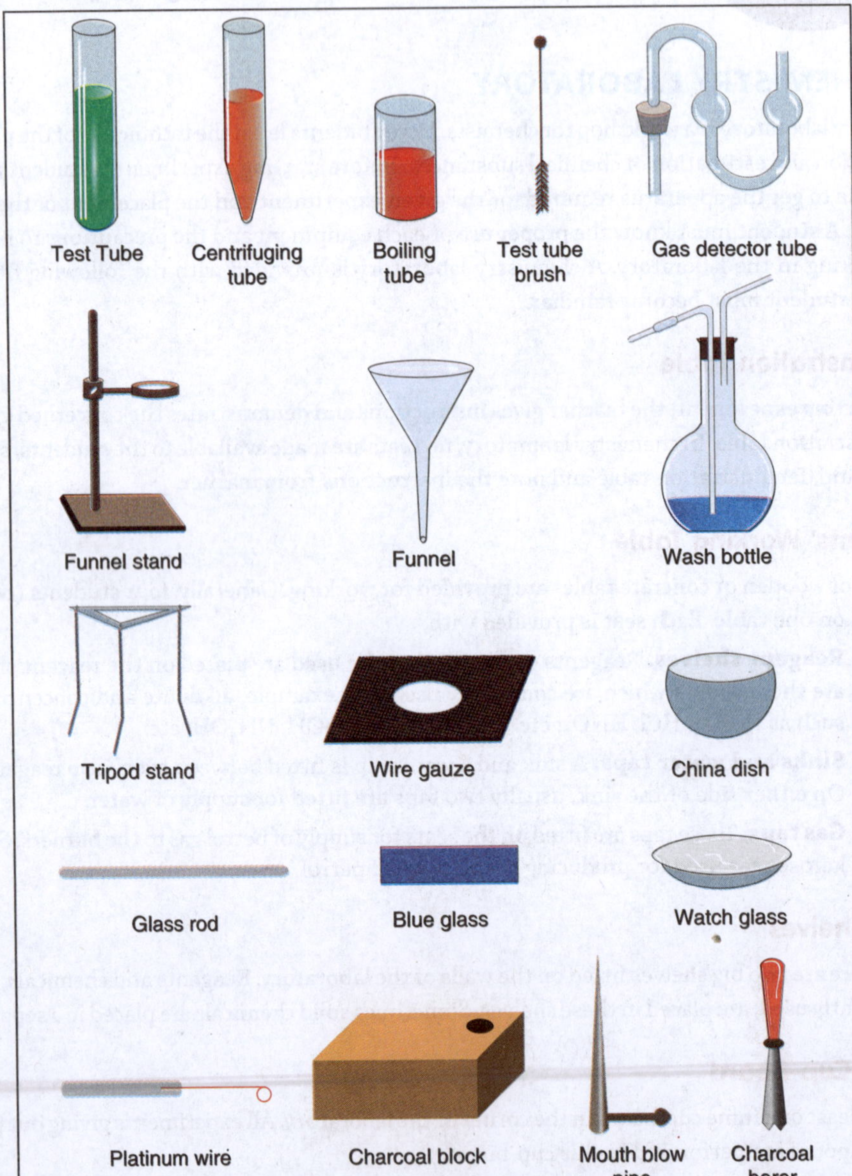

Fig. 1.1. Apparatus used in chemistry laboratory.

1. **Beakers.** Beakers of different sizes such as 150 mls, 200 mls made of soft glass or corning glass. Beakers are used for taking various liquids.
2. **Test Tubes.** Test tubes of different sizes are available. Small test tubes used for salt analysis known as centrifuging tubes and boiling tubes are also available.
3. **Conical Flask.** It is used in volumetric analysis for carrying out titration.
4. **Funnel.** It is used for filtration and for pouring solutions from one container to the other.
5. **Measuring Flask.** It is used in quantitative analysis when we have to prepare a solution with a particular volume. There are flasks of 50 ml, 100 ml and 250 ml capacity. There is a mark on the stem of the flask upto which the liquid is taken to complete the volume.
6. **Glass-Rod.** It is used for stirring purposes. It is also used as an aid for transferring the liquid into the funnel.
7. **China Dish.** It is a small vessel made of porcelain. It is used in crystallisation, for concentrating a solution.
8. **Wire Gauze.** It is placed above the flame of the burner so that the glass vessel being heated does not touch the flame directly and hence is prevented from breaking.
9. **Tripod Stand.** It is used for supporting a china dish or a beaker so that it can be heated from below.

Other apparatus with which a student must familiarize are *test tube holder, test tube brush, crucible tongs, spatula, watch glass, clamp stand, burette, pipette, water bath, sand bath and centrifugal machine*.

1.03. BUNSEN BURNER

It is a common heating device used in laboratory and consists of following parts:

1. **Base,** made of cast iron. It keeps the burner in a stable upright position.
2. **Gas-inlet tube.** It fits horizontally into the side of the base and can be connected to the gas tap through a rubber tube.
3. **Nipple,** made of brass rod and has a fine pin-hole running through it. At its lower end, the nipple is screwed into the base. At the upper end, it carries the burner base.

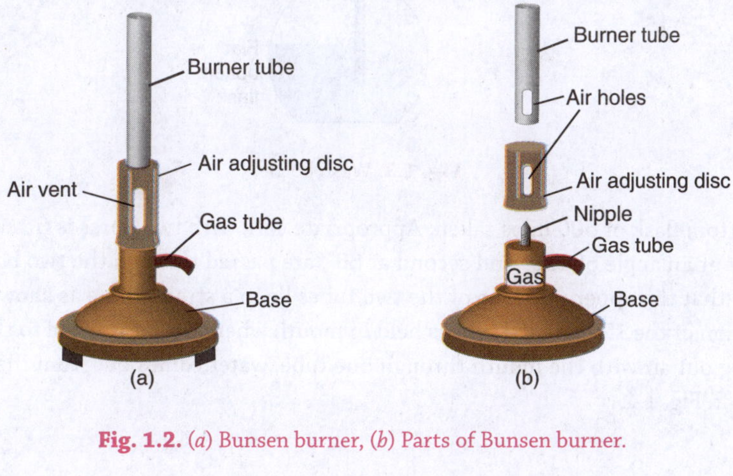

Fig. 1.2. (*a*) Bunsen burner, (*b*) Parts of Bunsen burner.

4. **Burner tube,** a metallic tube with two opposite air holes near its lower end. It is screwed to the nipple and carries the air regulator.
5. **Air regulator,** is a metallic ring that loosely fits on the lower end of the burner tube. It is pierced with two holes that exactly correspond to the two air holes of the burner tube. It can be rotated to regulate supply of air into the burner tube by partially or wholly closing the air holes.

Working of Burner

The rubber tubing is connected to the gas tap and the burner is lighted. As the gas escapes through the nipple, there is a fall of pressure. As a result of which air is sucked in through the air holes. The mixture of air and combustible gas burns at the top with a flame. Depending upon the quantity of air mixed, flame can be *luminous* or *non-luminous*.

Oxidising flame or non-luminous zone is hottest. It is this portion that should be used for the purpose of heating. Luminous zone is the brightest part of the flame. It is reducing in character and is used for reducing process, such as in charcoal cavity test, match stick test and borax bead test of some radicals.

1.04. WASH BOTTLE

A wash bottle is a container of distilled water with the help of which a fine stream of water can be obtained for washing the precipitate and for other purposes. It has the shape as shown in Fig. 1.3.

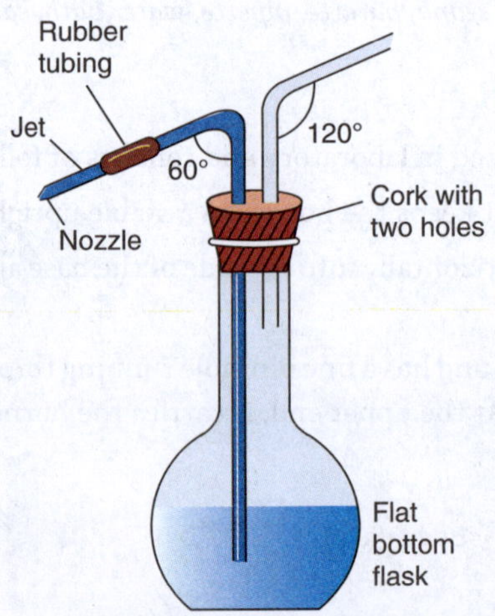

Fig. 1.3. Wash bottle.

A flat-bottom flask of 500 ml is taken. Appropriate cork with two bores is fitted into it. The two tubes, one bent at an angle of 120° and second at 60°, are passed through the two bores. This is done in a manner so that the upper portions of the two tubes lie in a straight line as shown in the Fig. 1.3. The upper portion of the 120° angled tube is held in mouth whereas a jet is fitted to the tube angled at 60°. On blowing out air with the mouth through one tube, water comes out from the other tube with force as shown in Fig. 1.4.

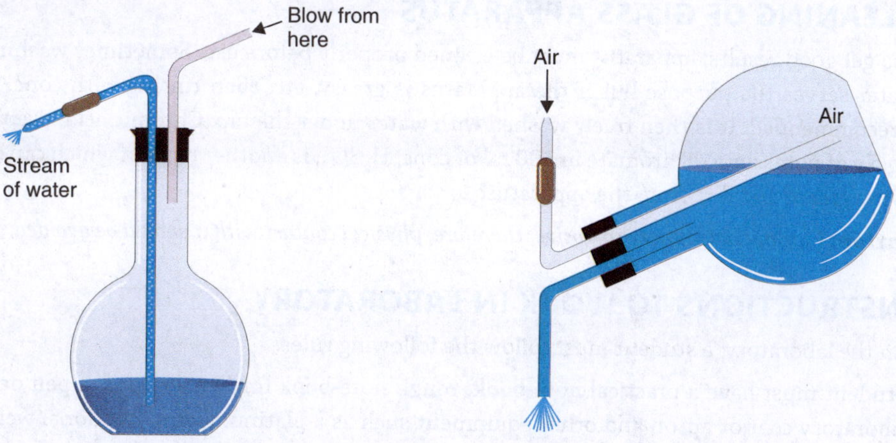

(a) For getting stream of water. (b) For getting a little larger amount of water.

Fig. 1.4. Use of a wash bottle.

PRECAUTIONS

1. The edges must be rounded off.
2. The longer arm of the tube bent at 60° should be only very slightly above the bottom of the flask so that it can be used even when it contains only a small amount of water.
3. All connections must be air tight.

Polythene Wash Bottle

Now-a-days polythene wash bottles are preferably used in the laboratory. It consists of flexible plastic material bottle, fitted with a plastic tubing having a jet at its outer end. On squeezing the bottle fine stream of water comes out of the jet. It can be used to give washings or to remove the precipitates from a beaker etc.

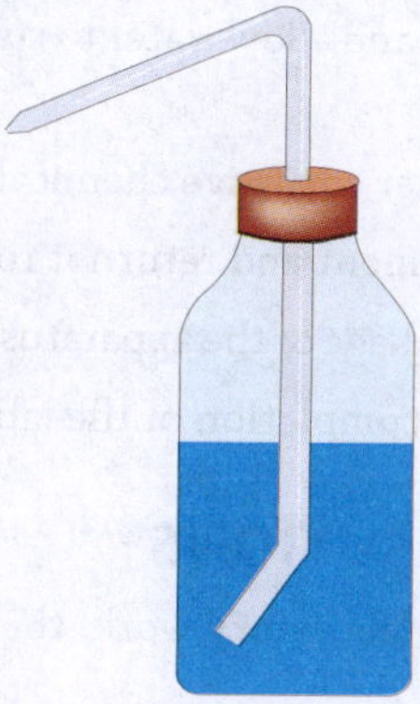

Fig. 1.5. Polythene wash bottle.

1.05. CLEANING OF GLASS APPARATUS

In order to get good results, apparatus must be cleaned properly before use. Sometimes washing with simple water serves the purpose but if the apparatus is greasy, etc. then rinsing with conc. HCl or HNO_3 is recommended. It is then freely washed with water under the tap. Chromic acid, prepared by dissolving 5 g of potassium dichromate in 100 ml of conc. H_2SO_4, is another reagent which can be used for removing grease and dirt from the apparatus.

Caution: *Chromic acid is very corrosive, therefore, physical contact with it should be avoided.*

1.06. INSTRUCTIONS TO WORK IN LABORATORY

To work in the laboratory, a student must follow the following rules:

1. A student must have a practical note-book, rough note-book for instructions, a pen or pencil, a laboratory coat or apron and other equipment such as a platinum wire, fractional weights as required.
2. Always come prepared for the experiment. This will help to understand the experiment better.
3. Always listen to the teacher's instructions carefully and note down the important points and precautions to be followed.
4. After the instructions, collect the apparatus from the laboratory assistant in queue.
5. Thoroughly clean the apparatus to be used.
6. Do only the experiments assigned, unallotted experiments should not be done.
7. Do your experiment honestly without caring for the final result. Record the observations on a rough note-book instead of writing on the pieces of paper.
8. Plan your work so as to finish in the stipulated time.
9. Be economical with the reagents. Only small quantities of the reagents are to be used. Use of reagents in excess not only leads to wastage of chemicals but also causes damage to the environment by polluting soil, water and air.
10. Handle the glass apparatus very carefully. In case of any breakage, report it to your teacher at once.
11. Dispose of all waste liquids in the sink and allow water to run for sometime by opening the water tap.
12. Keep your seat clean. If an acid or other corrosive chemical is spilled, wash it off with water.
13. Clean your apparatus after the experiment and return it to the laboratory assistant.
14. In case of any injury or accident or breakage of the apparatus, report it to the teacher immediately.
15. Wash your hands with soap after the completion of the laboratory work.

1.07. SOME IMPORTANT PRECAUTIONS

To avoid unnecessary risk or injury during laboratory work, the students are advised to observe the following precautions:

1. Always use an apron, an eye protector and hand gloves while working in the chemistry laboratory.
2. Do not touch any chemical with hand as some of them may be corrosive.

3. Before using any reagent or a chemical, read the label on the bottle carefully. Never use unlabelled reagent.
4. Do not put any object into the reagent bottle.
5. Do not bring inflammable liquids such as alcohol, ether near the flame.
6. Always pour acid into water for dilution. Never add water to acid.
7. Do not use cracked glass apparatus, such as beakers, for heating purposes.
8. Be careful while heating the test tube. The test tube should never point towards yourself or towards your neighbours while heating or adding a reagent.
9. Do not heat beakers or china dish directly on flame. Always make use of wire gauge.
10. Be careful in smelling chemicals or vapours. Always fan the vapours gently towards your nose (Fig. 1.6).

Fig. 1.6. Testing odours. Fan the vapour gently towards the nose.

1.08. PRACTICAL NOTEBOOK

All the experiments that are conducted in the laboratory are recorded in a practical notebook. It is compilation of whole work done by the student, so it must be well maintained, protected from mechanical and chemical damage. For keeping upto date record of experiments, following points should be kept in mind:

1. The name of the experiment should be entered along with the date of carrying out that experiment.
2. Requirements should be mentioned next to the title given.
3. Theory and principle of the experiment should be given in precise manner.
4. This should be followed by procedure in which experiment is to be conducted. Then a summary of precautions to be taken care are mentioned. Finally mention the general calculations for the experiment.

If we make a table of the points to be written on left hand and right hand side of the notebook, it will look somewhat like the one given as follows.

Left hand side	*Right hand side*
Date Name of experiment Diagram Chemical equation Observations Calculations	Date Name of experiment Theory Procedure General calculations Precautions

Keep following points in consideration regarding your practical notebook:

1. Do not tear pages from notebook.
2. Do not over write if a mistake has been committed in recording, put a line over it and write the correct word or figure.
3. Page mark your notebook.
4. Complete the index, indicating the experiment, its serial number, page number on which it is written.
5. Keep your notebook neat and tidy and covered with brown paper.

1.09. FIRST AID EMERGENCY TREATMENT IN THE LABORATORY

A chemistry laboratory encompasses different types of chemicals, apparatus. Any lack of attention on the part of student may cause accident. Accidents may occur by chance also. In any case prompt action should be taken to give first aid to the victim and then should be hospitalised if the need be. The probable accidents and their first aid emergency treatment are given below :

Type of accident	*First aid emergency treatment*
1. Burns:	
(*i*) Burn by dry heat (*i.e.*, flame, hot object etc.)	(*i*) Apply burnol or sarson (mustard) oil.
(*ii*) Burns causing blisters. **Caution:** *Heat burns should never be washed.*	(*ii*) Apply burnol at once.
(*iii*) Acid burns	(*iii*) Wash freely with water, then with 1% acetic acid and again with water. Dry the skin and apply burnol.
(*iv*) Bromine burns	(*iv*) Wash liberally with 2% NH_3 solution and then rub glycerine. Wipe off glycerine after some time and apply burnol.
2. Cuts:	
(*i*) Minor cuts	(*i*) Allow to bleed for a few seconds. Remove the glass piece if any. Apply a little methylated spirit and cover with a piece of cotton. Alternatively apply $FeCl_3$ solution to stop bleeding.
(*ii*) Serious cuts	(*ii*) Apply pressure above the cut to stop bleeding. Call for the doctor.

Type of accident	First aid emergency treatment
3. Eye Accidents:	
(i) Acid in eye	(i) Wash thoroughly with water, then with 1% sodium bicarbonate (Na_2CO_3) solution and then with water again.
(ii) Alkali in eye	(ii) Wash thoroughly with water then with 1% boric acid solution.
4. Poisons:	
(i) Poisons not swallowed	(i) Spit out immediately. Wash the mouth with water.
(ii) Acid swallowed	(ii) Drink lot of water. Drink lime water. No emetic should be taken.
(iii) Caustic alkalies swallowed	(iii) Drink lot of water. Drink a glass of lemon or orange juice. No emetic should be taken.
(iv) Inhalation of gases like Cl_2, SO_2, Br_2 etc. causing suffocation.	(iv) Loosen the clothes at the neck. Go in the open air. Inhale dilute vapours of ammonia or gargle with sodium bicarbonate solution.
5. Fire:	
(i) Clothes catch fire	(i) Do not run. Wrap with a blanket. Lie down on the floor and roll.
(ii) Beaker containing inflammable liquid catches fire	(ii) Cover the beaker with duster or damp cloth.

Go to the doctor after getting first aid.

VIVA QUESTIONS WITH ANSWERS

Q. 1. Why is a Bunsen burner provided with air holes?

Ans. To regulate the supply of air.

Q. 2. What type of flame would you use for general heating purpose?

Ans. A non-luminous oxidising flame as it gives maximum heat due to complete combustion of hydrocarbons.

Q. 3. What is the use of a fume cup-board?

Ans. It is used to perform those experiments which involve the release of poisonous gases or vapours.

Q. 4. What is the use of wash bottle?

Ans. It is used for getting a thin stream of water required for washing or transferring a precipitate.

Q. 5. What first aid is necessary when an acid gets into an eye while working in the laboratory?

Ans. The injured eye should be washed freely with water and then 1% solution of sodium bicarbonate.

CHAPTER 2

Surface Chemistry

Thomas Graham (1861) studied the process of diffusion of dissolved substances through a parchment paper or an animal membrane. He divided substances into two classes: (*i*) *crystalloids* and (*ii*) *colloids*. Substances like sugar, urea, common salt, etc. which readily passed through the membrane while in the dissolved state were called **crystalloids.** Substances like starch, glue, gelatine, etc., which in the dissolved state either do not pass at all or pass through very slowly are called **colloids.**

This arbitrary division of substances into crystalloids and colloids was soon proved to be wrong since a crystalloid could behave as a colloid under different conditions and *vice versa*. For example, common salt, a typical crystalloid in an aqueous solution, behaves as a colloid in the benzene medium, while soap, a typical colloid in water, behaves as a crystalloid in alcohol. A '*colloidal substance*', therefore, does not represent a separate class of substances. We now speak of the *colloidal state* of matter into which every substance can be obtained by a suitable method. The nature of the substance whether colloid or crystalloid depends upon size of the solute particles. When the size of the solute particles lies between 1 to 1000 nm, it behaves as a colloid. On the other hand, if size of solute particles is greater than 1000 nm, it exists as suspension and if particle size is less than 1 nm, it exists as true solution and behaves like a crystalloid. Thus, colloid is not a substance but is a particular state of the substance which depends upon size of its particles. The colloidal state is intermediate state between true solution and suspension.

It may be noted that a colloidal solution is heterogeneous in nature and always consists of at least two phases; the *disperse phase* and the *dispersion medium*.

Disperse phase. *It is the component present in small proportion and consists of particles of colloidal dimensions (1–1000 nm).*

Dispersion medium. *The medium in which colloidal particles are dispersed* is called dispersion medium. In a colloidal solution of sulphur in water, sulphur particles constitute disperse phase and water constitutes dispersion medium.

The two phases, namely dispersed and dispersion can be solid, liquid or a gas. Thus, different types of colloidal solutions are possible depending upon the physical state of the two phases. It should be borne in mind that gases between themselves cannot form a colloid due to their property of diffusion to give homogeneous mixtures.

2.01. LYOPHILIC AND LYOPHOBIC SOLS

Depending upon the interaction between the disperse phase and the dispersion medium, the colloidal solutions are classified into two types:

1. *Lyophilic sols;* and
2. *Lyophobic sols.*

1. Lyophilic Sols

In this type of colloidal solutions, the disperse phase has great affinity for the dispersion medium. In such colloids, the dispersed phase does not get easily precipitated and the sols are quite stable. The solids obtained after evaporation may be reconverted to the sol state by simply agitating them with the dispersion medium. Examples of lyophilic sols include sols of gum, gelatine, starch, proteins and certain polymers in organic solvents. Such sols are called **reversible sols.** If water is the dispersion medium, these are called *hydrophilic sols*.

2. Lyophobic Sols

In this type of sols, disperse phase has little affinity for the dispersion medium. These sols are relatively less stable than lyophilic sols. These sols are easily precipitated (or coagulated) on the addition of small amounts of electrolytes, by heating or by shaking. These are *irreversible* as their precipitated mass cannot be brought back into the colloidal state by simply shaking them up with the dispersion medium. Examples of lyophobic sols include sols of metals, and their insoluble compounds like sulphides and oxides. They need stabilizing substances for preservation. If water is the dispersion medium, these are known as *hydrophobic sols*.

EXPERIMENT 2.1

AIM: ***To prepare colloidal solution (sol) of starch.***

THEORY

Starch forms a *lyophilic* sol when water is used as the dispersion medium. The formation of sol is accelerated by heating. The starch sol can be prepared by heating starch and water at about 100°C. It is quite stable and is not affected by the presence of any electrolytic impurity.

APPARATUS

Beakers (250 ml and 50 ml), glass rod, funnel, filter-paper, pestle and mortar, tripod stand, wire-gauze and burner.

MATERIALS REQUIRED

Soluble starch (500 mg) and distilled water.

PROCEDURE

1. Take 500 mg of starch in a mortar and add few ml of distilled water.
2. Grind the starch to make a thin paste and transfer this paste to a 50 ml beaker.
3. Take about 100 ml of distilled water in a 250 ml beaker and heat the beaker so that water starts boiling.
4. Pour the paste slowly with stirring into boiling water in the beaker (Fig. 2.1).
5. Continue boiling for about 10 minutes and then allow the beaker to cool.
6. Filter the contents of the beaker through a filter-paper, fixed in a funnel.

   Label the filtrate **'Starch Sol'**.

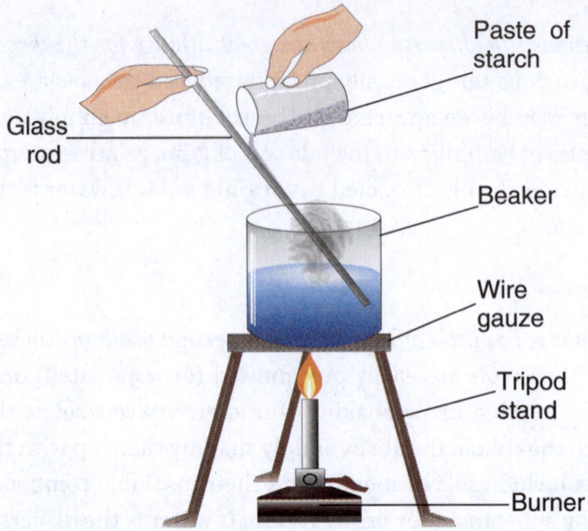

Fig. 2.1. Preparation of starch sol.

PRECAUTIONS

1. The apparatus used for preparing sol should be properly cleaned.
2. Distilled water should be used for preparing sols in water.
3. Starch should be converted into a fine paste before adding to boiling water.
4. Starch paste should be added in a thin stream to boiling water.
5. Constant stirring of the contents is necessary during the preparation of the sol.

EXPERIMENT 2.2

AIM: ***To prepare a colloidal solution of gum.***

Hint: This experiment can be performed in a similar way as explained in Expt. 2.1 with the exception that instead of boiling water, warm water is to be used since gum is quite soluble even in warm water.

EXPERIMENT 2.3

AIM: ***To prepare colloidal solution (or sol) of egg albumin.***

THEORY

Egg albumin which is obtained from eggs forms lyophilic sol with cold water. The sol is quite stable and is not affected by the presence of traces of impurities.

APPARATUS

Same as in Experiment 2.1.

MATERIALS REQUIRED

An egg and distilled water.

PROCEDURE

1. Break the outer shell of the egg by striking it with a glass rod and collect its colourless liquid along with yellow yolk. Decant the colourless liquid into another beaker. This colourless liquid is known as *egg albumin*.
2. Prepare 100 ml of 5% (*w/v*) solution of sodium chloride in a 250 ml beaker. To this solution add egg albumin in small portions with constant stirring. This process should take 15–20 minutes.
3. Filter the contents of the beaker through a filter paper, fixed in a funnel, and collect the filtrate. Label this filtrate as **'Egg-albumin Sol'**.

PRECAUTIONS

1. The apparatus used for preparing the sol should be absolutely clean.
2. Distilled water should be used for preparing the sol.
3. **Egg albumin sol is prepared at room temperature** because in hot solution the precipitation of egg albumin takes place.
4. The yellow yolk should be separated completely from the egg albumin before using the latter in the experiment.
5. Addition of egg albumin should be done very slowly and with constant stirring so as to disperse the colloidal particles well in solution.

EXPERIMENT 2.4

AIM: ***To prepare ferric hydroxide, [Fe(OH)$_3$] sol.***

THEORY

Ferric hydroxide forms a **lyophobic sol.** The substances such as metal hydroxides or sulphides which are insoluble and do not readily give colloidal solutions on treatment with water are called **lyophobic colloids.**

Ferric hydroxide sol is prepared by the hydrolysis of ferric chloride with boiling distilled water. The reaction that takes place can be represented as

$$\underset{\text{Ferric chloride}}{FeCl_3(aq)} + 3H_2O(l) \xrightarrow{\text{Boil}} \underset{\text{Red sol}}{Fe(OH)_3(s)} + 3HCl(aq)$$

The hydrolysis reaction produces insoluble ferric hydroxide particles which undergo *agglomerisation* to yield bigger particles of colloidal dimensions. These particles adsorb Fe^{3+} ions preferentially from the solution to give positive charge to the sol particles. Stability of the sol is due to the charge on the sol particles. *Hydrochloric acid which is produced during hydrolysis destabilizes the sol and hence it must be removed from the sol by dialysis process, otherwise sol will not be stable.*

APPARATUS

Conical flask (250 ml), beaker (250 ml), a boiling-tube, glass-rod, funnel, round-bottom flask, iron stand with a clamp, wire-gauze, tripod-stand, burner and a burette or a dropper.

MATERIALS REQUIRED

2% solution of ferric chloride (prepared by dissolving 2 g of pure $FeCl_3$ in 100 ml distilled water) and distilled water.

PROCEDURE

1. Take a 250 ml conical flask and clean it by steaming-out process as shown in Fig. 2.2.

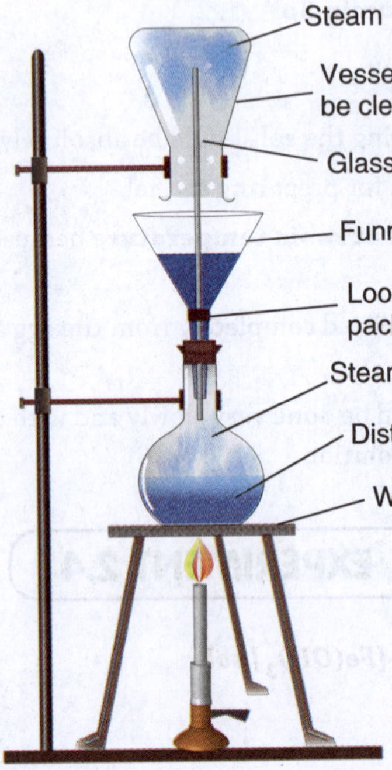

Fig. 2.2. Steaming-out process for cleaning conical flask.

2. To this cleaned flask, add 100 ml of distilled water and heat it to boil by placing the flask on a wire-gauze.
3. Add ferric chloride solution dropwise (by the use of a burette or a dropper) to the boiling water.
4. Continue heating until deep red or brown solution of ferric hydroxide is obtained. Replace the water lost by evaporation during boiling at regular intervals.
5. Keep the contents of conical flask undisturbed for sometime at room temperature. Label the solution as **"Ferric hydroxide Sol"**.

PRECAUTIONS

1. Since ferric hydroxide sol is affected by impurities, the apparatus required for the preparation of sol should be thoroughly cleaned by steaming-out process.

2. Add ferric chloride solution dropwise.
3. Heating is continued till the desired sol is obtained.
4. Hydrochloric acid formed as a result of hydrolysis of ferric chloride is removed by dialysis process otherwise it would destablise the sol.

EXPERIMENT 2.5

AIM: ***To prepare aluminium hydroxide [Al(OH)$_3$] sol.***

THEORY

Aluminium hydroxide sol is *hydrophobic* in nature. It is obtained by hydrolysis of aluminium chloride

$$AlCl_3 + 3H_2O \longrightarrow \underset{\text{(Sol)}}{Al(OH)_3} + 3HCl$$

Dialysis is done to remove hydrochloric acid (produced as a result of hydrolysis of aluminium chloride) because aluminium hydroxide sol is affected by the presence of ionic impurities.

APPARATUS

Same as in Experiment 2.4.

MATERIALS

Aluminium chloride (2% solution) and distilled water.

PROCEDURE

Same as in Experiment 2.4.

EXPERIMENT 2.6

AIM: ***To prepare colloidal solution of arsenious sulphide, [As$_2$S$_3$].***

THEORY

Arsenious sulpide, As_2S_3 is a *lyophobic colloid*. It is obtained by the hydrolysis of arsenious oxide (As_2O_3) with boiling distilled water, followed by passing H_2S gas through the solution obtained.

$$As_2O_3 + 3H_2O \xrightarrow{\text{Boil}} 2As(OH)_3$$

$$2As(OH)_3 + 3H_2S \longrightarrow \underset{\substack{\text{(Light}\\\text{yellow sol)}}}{As_2S_3} + 6H_2O$$

In the colloidal solution of arsenious sulphide, each particle is surrounded by HS^- ions, produced by the dissociation of H_2S. This sulphide ion layer is further surrounded by the counter ion layer of H^+ ions.

APPARATUS

Conical flasks (250 ml), beaker (250 ml), round-bottom flask (500 ml), glass-rod, funnel, glass tubing, filter-paper, tripod stand, wire-gauze, iron stand with clamp, burner, etc.

MATERIALS REQUIRED

Solid arsenious oxide, H_2S gas and distilled water.

PROCEDURE

1. Clean a conical flask (250 ml) by the use of steaming-out process.
2. To this cleaned flask, add 0.2 g of As_2O_3 solid and add 100 ml of distilled water.
3. Boil the solution for about 10 minutes (Fig. 2.3).

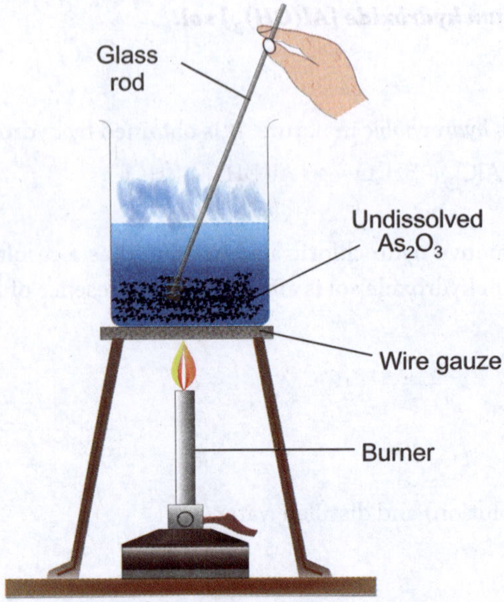

Fig. 2.3

4. Filter the above hot solution through flutted filter paper and receive the filtrate in another beaker (Fig. 2.4).

Fig. 2.4

5. Pass a slow current of H_2S into As_2O_3 solution as shown in Fig. 2.5.

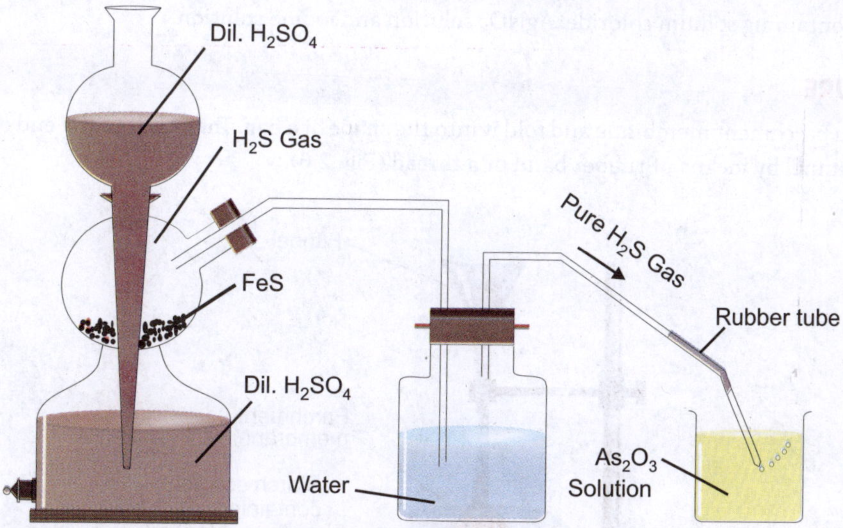

Fig. 2.5. Preparation of As_2S_3 sol.

The solution develops a yellow colour due to formation of As_2S_3. Continue passing H_2S till the intensity of colour does not change further.

6. Expel excess of H_2S gas from the sol by boiling the sol till the escaping gas does not turn lead acetate paper black.
7. Filter the solution through flutted-filter paper and collect the bright yellow filtrate in a dry conical flask and cork it. Label it as **"Arsenious Sulphide Sol"**.

PRECAUTIONS

1. Use cleaned apparatus since As_2S_3 sol is affected by even traces of impurities.
2. Handle arsenious oxide with care since it is **highly poisonous.** Wash your hands immediately everytime after handling this chemical. While doing this experiment do not eat or drink anything.

EXPERIMENT 2.7

AIM: ***To study the dialysis of starch sol containing sodium chloride through a cellophane or parchment paper.***

THEORY

The purification of sols by dialysis is based upon the fact that while the colloidal particles cannot pass through cellophane or parchment membrane, the ions of an electrolyte can readily do so.

APPARATUS

A 400 ml beaker, a funnel with a long stem, cellophane or parchment membrane, dropper, test tubes and iron-stand.

MATERIALS REQUIRED

Starch sol containing sodium chloride, $AgNO_3$ solution and iodine solution.

PROCEDURE

1. Take a parchment membrane and fold it into the shape of a bag. Then tie it to the end of the stem of a funnel by means of rubber band or a thread (Fig. 2.6).

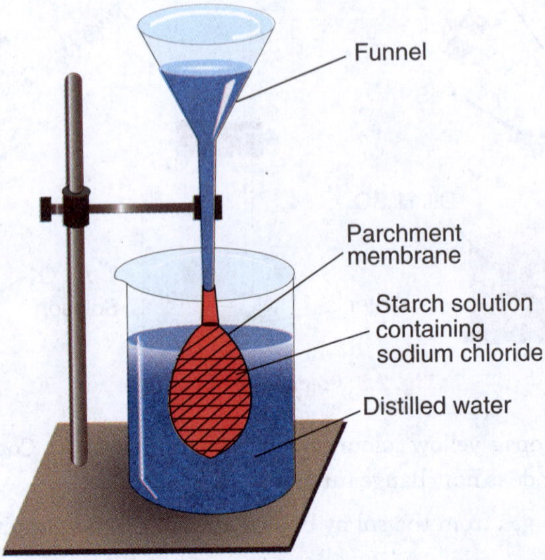

Fig. 2.6. Purification of starch sol by dialysis.

2. Add the given starch sol containing sodium chloride into the parchment bag through the funnel till two-third of the bag is full.
3. Take a 400 ml beaker and fill it three-fourth with distilled water. Place it over an iron-stand, dip the parchment bag into distilled water and fix the funnel in position by means of a clamp.
4. Allow to it stand for about half an hour.
5. Then, withdraw about 1 ml of water from the beaker with the help of a dropper and transfer it to a test tube. Add to it a few drops of iodine solution. *No blue colour appears*. This indicates the *absence of starch* in water. Thus, it follows that starch molecules do not diffuse through parchment paper.
6. Now withdraw another 1 ml of water from the beaker and transfer it to another test tube. Add to it a few drops of $AgNO_3$ solution. *A white ppt. of AgCl is produced immediately*. This shows presence of chloride ions and hence sodium chloride in water. It follows that Na^+ and Cl^- diffuse through the parchment paper. As Na^+ and Cl^- diffuse out of the starch sol, it gets free from the ions gradually.
7. In order to check whether sodium chloride is completely removed or not replace the water in the beaker by fresh distilled water and again place the parchment bag containing sol in it. After about 10 minutes, test for the presence of Cl^- ions. If the Cl^- ions are absent dialysis is complete, otherwise the sol still contains Cl^- ions and therefore the dialysis should be continued.

PRECAUTIONS

1. Fill only two-third of the cellophane/parchment bag with sol.
2. There should be no leakage of sol from the bag into the beaker.
3. Use distilled water for dialysis.

2.02. COAGULATION OR PRECIPITATION OF COLLOIDAL SOLUTIONS

Presence of small concentrations of appropriate electrolytes is necessary to stabilize the colloidal solutions. However, if the electrolytes are present in higher concentration then the ions combine with charged colloidal particles and neutralize them. Now these colloidal particles may unite together to form bigger particles which are then precipitated. *The precipitation of a colloid through induced aggregation by the addition of some suitable electrolyte is called* **coagulation** or **flocculation.**

The coagulation of a colloidal solution by an electrolyte does not take place until the added electroyte has certain minimum concentration in the solution. *The minimum concentration of an electrolyte in millimoles per litre of the mixed solution, required to cause coagulation of a particular sol* is called the **coagulation** or **precipitation value of the electrolyte for the sol.**

Different electrolytes have different coagulation values. The coagulating behaviour of various electrolytes was studied in detail by **Hardy** and **Schulze.** They observed that:

(*i*) *The ions carrying charge opposite to that of sol particles are effective in causing the coagulation of the sol.*

(*ii*) *Coagulation power of an electrolyte is directly proportional to the valency of ion causing coagulation.*

Thus, for the coagulation of sols carrying negative charge (like As_2S_3 sol), Al^{3+} ions are more effective than Ba^{2+} or Na^+ ions. Similarly, for the coagulation of sols carrying positive charge (such as $Fe(OH)_3$ sol) PO_4^{3-} ions are more efficient than SO_4^{2-} or Cl^- ions. The two observations given above are collectively called **Hardy-Schulze rule.**

Coagulation of colloidal solutions can also be achieved by the following methods:

***(i)* By mutual precipitation.** When two oppositely charged sols (such as of $Fe(OH)_3$ and As_2S_3) are mixed in equi-molar proportions, they neutralize each other and get coagulated.

***(ii)* By electrophoresis.** We know that during electrophoresis the sol particles move towards the oppositely charged electrodes. If the process is carried for a long time, the particles will touch the electrode, lose their charge and get coagulated.

***(iii)* By repeated dialysis.** The stability of colloidal sols is due to the presence of a small amount of electrolyte. If the electrolyte is completely removed by repeated dialysis, the sol will get coagulated.

***(iv)* By heating.** The sol may be coagulated even by simple heating.

EXPERIMENT 2.8

AIM: ***Compare the precipitation values of sodium chloride, barium chloride and aluminium chloride for arsenious sulphide sol.***

THEORY

Arsenious sulphide sol consists of colloidal particles of arsenious sulphide, As_2S_3, dispersed in water. It is a negatively charged sol. It will be precipitated by positively charged ions.

However, the precipitation value of the cation will depend upon the valency of the cation. Thus univalent cations such as sodium ion (Na^+) have much less precipitation power and, therefore, in their case large amount of salt will have to be added to precipitate the arsenious sulphide sol. On the other hand, bivalent cations such as barium ions (Ba^{2+}) have much greater precipitation power and a lesser amount of such salts will be needed for precipitating the arsenious sulphide sol. The trivalent cations such as aluminium ions (Al^{3+}) have still greater precipitation power and still lesser amount of aluminium salt is needed for precipitating arsenious sulphide sol.

APPARATUS

Three conical flasks (100 ml), a burette and a beaker.

MATERIALS REQUIRED

Arsenious sulphide sol, 0.1 M sodium chloride solution. 0.01 M barium chloride solution, 0.001 M aluminium chloride solution.

PROCEDURE

1. Take three conical flasks and label them as 1, 2, 3.
2. To each conical flask transfer 20 ml of the arsenious sulphide sol.
3. To the sol in the conical flask No. 1 add 0.1 M NaCl solution drop by drop with the help of a burette. Shake gently after the addition of each drop. Continue the addition till a yellow precipitate of arsenious sulphide is just formed. Note the volume of sodium chloride solution required for precipitation of arsenious sulphide sol.
4. Now, to the sol in conical flask No. 2 add 0.01 M $BaCl_2$ solution dropwise. Measure the volume when a yellow precipitate of arsenious sulphide is just formed.
5. Similarly, to the sol in the conical flask No. 3 add 0.001 M. $AlCl_3$ solution dropwise. Measure the volume when a yellow precipitate of arsenious sulphide is first formed.

OBSERVATIONS

Conical flask	*Volume of As_2S_3 sol taken*	*Electrolyte solution added*	*Concentration of electrolyte*	*Volume of electrolyte solution added*	*Total volume*	*Precipitation value*
1.	20 ml	NaCl	0.1 M	x ml	$20 + x = V_1$ ml	$\frac{0.1 \times x \times 1000}{V_1}$
2.	20 ml	$BaCl_2$	0.01 M	y ml	$20 + y = V_2$ ml	$\frac{0.1 \times y \times 1000}{V_2}$
3.	20 ml	$AlCl_3$	0.001 M	z ml	$20 + z = V_3$ ml	$\frac{0.001 \times z \times 1000}{V_3}$

RESULT

The *precipitation values* of NaCl, $BaCl_2$ and $AlCl_3$ for As_2S_3 sol are in the order

$$NaCl > BaCl_2 > AlCl_3$$

On the other hand, the coagulating or the *precipitating powers* of these electrolytes are in the order

$$AlCl_3 > BaCl_2 > NaCl.$$

PRECAUTIONS

1. The apparatus to be used should be cleaned thoroughly.
2. Look vertically down the conical flask for detecting the start of the precipitation.
3. Mixing of sol and electrolyte solution should be done by gentle inverting the corked test-tube. **Do not shake them vigorously.**

2.03. EMULSIONS AND EMULSIFYING AGENTS

Emulsions are colloidal solutions in which disperse phase as well as dispersion medium are liquids. Emulsions can be broadly classified into two types:

1. **Oil in water emulsions.** In this type of emulsions, oil acts as disperse phase and water acts as dispersion medium. Some examples of this type of emulsions are *milk, vanishing cream*, etc.
2. **Water in oil emulsions.** In this type of emulsions, water acts as disperse phase and oil acts as dispersion medium. For example, *butter, cod-liver oil*, etc.

Preparation of Emulsions

The process of making an emulsion is known as *emulsification*. Emulsions may be obtained by vigorously agitating a mixture of both the liquids. But this gives an unstable emulsion. The dispersed drops at once come together and form a separate layer. To stabilize an emulsion, the addition of a small quantity of the third substance known **as emulsifying agent** or **emulsifier** is essential. *Soaps* and *detergents* are most frequently used as emulsifiers. They coat the drops of the emulsion and check them from coming together and the emulsion is thus stabilized. The other common stabilizing agents are *proteins, gum* and *agar-agar*.

Demulsification

The decomposition of an emulsion into its constituent liquids is called demulsification. The various techniques applied for demulsification are freezing, boiling, filtration, centrifugation, electrostatic precipitation or chemical methods which destroy the emulsifying agents. For example, cream is separated from milk by centrifugation.

Applications of Emulsions

1. Washing action of soaps and detergents is due to the emulsification of grease and taking it away in the water along with dirt and dust present on grease.
2. A wide variety of pharmaceutical preparations are emulsions. For example, emulsion of cod liver oil. These emulsified oils are easily acted upon by the digestive juices in the stomach and hence are readily digested.

3. The disinfectants such as phenyl, dettol, and lysol give emulsions of the oil-in-water type when mixed with water.
4. Emulsions play an important role in industry. The metal ores are concentrated by froth-floatation process which involves the treatment of the pulverized ore in emulsion of pine oil.

EXPERIMENT 2.9

AIM: ***To study the effectiveness of different common oils (castor oil, cotton seed oil, coconut oil, kerosene oil, mustard oil) in forming emulsions.***

THEORY

Depending upon the physical properties and the chemical composition of the oil, some oils form emulsions readily with water whereas many others form with great difficulty.

APPARATUS

Five stoppered bottles or boiling tubes, measuring cylinder, stop-watch or simple watch, 5 ml pipettes.

MATERIALS REQUIRED

Castor oil, cotton seed oil, coconut oil, kerosene oil, mustard oil and 1% soap solution or sodium oleate solution. (1 gm dissolved in 100 ml of distilled water).

PROCEDURE

1. Take five stoppered bottles and wash them with water. Label them as A, B, C, D and E.
2. Take 5 ml of each of the oils separately in bottles A, B, C, D and E.
3. Add 50 ml of distilled water to each bottle.
4. Take bottle A, stopper it and shake vigorously for one minute. Then allow it to stand.
5. Note the time taken for the two layers to separate out.
6. Similarly, take bottles B, C, D and E and note the time taken for the separation of two layers in each case.
7. Now add 10 drops of 1% soap solution or 1% sodium oleate solution to each of the five bottles and find out the time taken for the two layers to separate. Record the observations.

OBSERVATIONS

Bottle or Tube	***Name of the oil***	***Time taken for the separation of layers***	
		Without soap/detergent	***With soap/detergent***
A	Castor oil	— *s*	— *s*
B	Coconut oil	— *s*	— *s*
C	Mustard oil	— *s*	— *s*
D	Cotton seed oil	— *s*	— *s*
E	Kerosene oil	— *s*	— *s*

RESULT

It is clear from the above observations that ...... oil takes longest time to get separated from its emulsion and is rated 1 and ...... oil takes the minimum time and is rated 5. The decreasing order of stability or effectiveness is ......

PRECAUTIONS

1. Add equal number of drops of soap solution to all the bottles.
2. Each bottle should be shaken vigorously and for the same time.
3. The time should be recorded carefully. Start the stopwatch immediately after shaking is stopped and stop it immediately when the two layers just separate.

EXPERIMENT 2.10

AIM: ***To compare the effectiveness of a number of emulsifying agents in forming emulsions.***

THEORY

Different emulsifying agents have different capacities for emulsifying a given oil. An emulsifying agent lowers the interfacial tension between water and oil and gets concentrated at the surface between two liquids. Due to the reduced interfacial tension, the tiny droplets of oil do not coalesce and thus the emulsions become stable. Since different emulsifying agents have different tendencies to lower the interfacial tensions, they have different capacities for emulsifying a given oil.

APPARATUS

Five stoppered bottles, measuring cylinder, stop-watch or simple watch, 5 ml pipettes.

MATERIALS REQUIRED

Castor oil, 1% solutions of sodium oleate, soap, detergent, gelatine and gum acacia.

PROCEDURE

1. Take five stoppered bottles and wash them with water and label them as A, B, C, D and E.
2. Take 5 ml of castor-oil in each of five bottles, A, B, C, D and E.
3. Add 50 ml of distilled water to each bottle.
4. Add 5 drops of sodium oleate solution to bottle A, shake it vigorously for one minute and allow it to stand.
5. Note the time taken for the two layers to separate out.
6. Similarly, take tubes, B, C, D and E and add 5 drops of soap solution, detergent solution, gelatine solution and gum acacia solution respectively to them. Shake vigorously for one minute and observe the time taken for the two layers to separate out in each case. Record the observations.

OBSERVATIONS

Volume of castor oil taken in each tube = 5 ml

Volume of distilled water added = 50 ml

Bottle or Tube	*Emulsifier added*	*Volume of emulsifier added*	*Time taken for the separation of two layers*
A	1% Sodium oleate solution	5 drops	— s
B	1% Soap solution	5 drops	— s
C	1% Detergent solution	5 drops	— s
D	1% Gelatine	5 drops	— s
E	1% Gum acacia	5 drops	— s

RESULT

It is clear from the above observations that the emulsifier......... when added causes the emulsion to take the longest time to break and is rated 1 while the emulsifier...... when added causes the emulsion to take the minimum time to break and is rated 5. The effectiveness of the given emulsifying agents is in the following order.........

PRECAUTIONS

Same as in experiment 2.9.

VIVA QUESTIONS WITH ANSWERS

Q. 1. What is a true solution? Give an example.

Ans. A homogeneous mixture of two substances, in which particles of both possess *molecular dimensions* (*i.e.*, 10^{-10} to 10^{-9} m). For example, a solution of common salt in water.

Q. 2. What is a colloidal solution? Give an example.

Ans. A colloidal solution is a *"two-phase heterogenous system in which a substance is distributed in colloidal state (particles having diameter between 1 to 1000 nm) in a medium"*. The particles of the dispersed substance (of colloidal size) are called dispersed phase; while the medium in which they are dispersed, is called *dispersion medium*. For example, milk, butter, smoke, etc.

Q. 3. What are lyophilic and lyophobic sols?

Ans. The sols in which colloidal particles have considerable affinity for the dispersion medium are termed as lyophilic sols. The sols in which colloidal particles have very little or no affinity for the dispersion medium are termed as lyophobic sols. For example, starch sol is a lyophilic sol whereas sulphur sol is a lyophobic sol.

Q. 4. What is the size of colloidal particles?

Ans. Between 1—1000 nm.

Q. 5. What is a sol and what is a gel?

Ans. When a colloidal solution appears as a fluid, it is called a sol but if it has a semi-solid appearance, it is known as a gel. In sol liquid is the dispersion medium and solid the disperse phase whereas in gel solid is the dispersion medium and liquid the disperse phase.

Q. 6. Give two examples of positively charged sols.

Ans. Ferric hydroxide sol and aluminium hydroxide sol.

Q. 7. Give an example of negatively charged sol.

Ans. Arsenious sulphide sol.

Q. 8. Why can't we prepare a colloidal solution of a gas in a gas?

Ans. The mixture of any two gases is always a homogeneous mixture and therefore does not satisfy the condition for the formation of a colloidal solution.

Q. 9. What is meant by the term dialysis?

Ans. The process of separating electrolytes from a colloid by means of diffusion of the former through an animal or vegetable membrane, is called *dialysis*.

Q. 10. What is the use of dialysis?

Ans. Dialysis is used for *purifying colloidal solution*. When a colloidal sol containing impurities is kept in a parchment bag, the electrolytes pass through the membrane while colloidal particles are retained by parchment bag.

Q. 11. How can we make dialysis fast?

Ans. By circulating hot water instead of cold water, in the container or by applying electric field across the dialyser.

Q. 12. How can a colloidal solution and a true solution of the same colour be distinguished from each other?

Ans. When a powerful beam of light is passed through true and colloidal solutions, each kept in glass vessel, then *only colloidal solution exhibits Tyndall effect* whereas true solution does not.

Q. 13. What is peptization?

Ans. The method of breaking down a precipitate into colloidal form by shaking it with the dispersion medium in the presence of an electrolyte, is called *peptization*. The electrolyte used for this purpose, is called *peptizing agent*.

Q. 14. What is coagulation?

Ans. Coagulation is a process of changing the colloidal particles into an insoluble precipitate, by inducing aggregation of colloidal particles.

Q. 15. Name a few methods of coagulating lyophillic colloids.

Ans. (*a*) Addition of electrolyte.

(*b*) Addition of a liquid in which the dispersion medium is soluble.

Q. 16. What is the reason for the coagulation of a lyophillic sol by addition of electrolyte?

Ans. Addition of excess of an electrolyte neutralizes the charge on colloidal particles which therefore unite to form bigger particles and get precipitated.

Q. 17. Name a few methods of coagulating lyophobic sols.

Ans. (*a*) By adding electrolytes.

(*b*) By boiling.

(*c*) By Electrophoresis.

Q. 18. What do you understand by coagulation value?

Ans. The minimum amount of the electrolyte in millimoles per litre of the combined solution required to cause coagulation is called the coagulation value of the electrolyte for a particular sol.

Q. 19. What is Hardy-Schulze rule?

Ans. **Hardy** and **Schulze rule** *states that the coagulating power of an electrolyte depends upon the valency of ion carrying charge opposite to that of the dispersed phase.*

Q. 20. Which one of the following electrolytes brings about the coagulation of As_2S_3 sol quickest and in the least concentration?

(*a*) NaCl

(*b*) $MgSO_4$

(*c*) $AlPO_4$.

Ans. $AlPO_4As_2S_3$ sol is negatively charged and is coagulated by adding positively charged ions. Al^{3+} has the greatest positive valence of all the electrolytes given. Hence, it is the most effective in causing coagulation.

Q. 21. What is the characteristic of dialysing membrane?

Ans. It allows only the electrolytes to pass through it, but does not allow the sol particles to pass through.

Q. 22. Give names of the two substances which are used for the preparation of dialysing membranes.

Ans. Cellophane and parchment.

Q. 23. What is an emulsion?

Ans. An emulsion is a colloidal system in which the dispersion medium as well as the dispersed phase are liquids.

Q. 24. How many types of emulsions are known?

Ans. Two, *oil-in-water* emulsion and *water-in-oil* emulsion.

Q. 25. What is oil-in-water emulsion? Give two examples.

Ans. Oil-in-water emulsion is that in which oil forms the dispersed phase and water the dispersion medium. For example, milk and vanishing cream.

Q. 26. What is water-in-oil emulsion? Give two examples.

Ans. Water-in-oil emulsion is that in which water forms the dispersed phase and oil forms the dispersion medium. For example, cold cream and cod-liver oil.

Q. 27. Differentiate between a solution and an emulsion.

Ans. A solution is a homogeneous mixture of two liquids whereas an emulsion is a heterogeneous mixture of two liquids.

Q. 28. What is an emulsifier?

Ans. Any substance used to stabilize an emulsion is known as emulsifier.

Q. 29. What is the role played by an emulsifier in the stability of emulsions?

Ans. An emulsifier lowers the interfacial tension between the two immiscible liquids. It envelopes the droplets of the dispersed phase and therefore they do not coalesce.

Q. 30. Is it possible to get an emulsion by mixing two miscible liquids?

Ans. No; it will lead to the formation of solution.

Q. 31. What is emulsification?

Ans. It is the process of getting an emulsion from its components.

Q. 32. What is demulsification?

Ans. It is the process of splitting an emulsion into its constituent liquids.

Q. 33. How would you differentiate between oil-in-water and water-in-oil emulsion?

Ans. (i) *By dye test.* To the emulsion add water soluble dye. If the droplets become coloured the emulsion is oil-in-water type and if the medium becomes coloured the emulsion is water-in-oil type.

(ii) *By dilution test.* To the emulsion add a few drops of oil. If the oil forms a separate layer, the emulsion is oil-in-water type, otherwise it is water-in-oil type.

CHAPTER 3 Chemical Kinetics

During our studies, we have come across so many reactions. These reactions proceed at different speeds. The speeds of reactions vary from very slow to very fast. For example, rusting of iron takes place very slowly whereas precipitation of silver chloride, on mixing the solutions of silver nitrate and sodium chloride, takes place at once. On the other hand, there are several reactions which proceed at measurable speeds. Inversion of sucrose and hydrolysis of starch are two such reactions :

$$\underset{\text{Cane sugar}}{C_{12}H_{22}O_{11}} + H_2O \longrightarrow \underset{\text{Glucose}}{C_6H_{12}O_6} + \underset{\text{Fructose}}{C_6H_{12}O_6}$$

$$\underset{\text{Starch}}{2(C_6H_{10}O_5)_n} + nH_2O \longrightarrow \underset{\text{Maltose}}{nC_{12}H_{22}O_{11}}$$

The rates of such reactions can be studied and experimentally determined. *The branch of chemistry which deals with the study of reaction rates and their mechanisms is called* **Chemical Kinetics.**

Before we take up study of any such reaction, let us re-capitulate about rates of reactions and factors influencing it.

In any chemical reaction, reactants are consumed and new products are formed. This means that we can measure the rate of a reaction in terms of the rate at which the reactants are consumed or the rate at which the products are obtained. **Rate of a reaction** is defined as *the rate of change in concentration of any of the reactants or products at a particular moment of time.*

3.01. FACTORS AFFECTING RATE OF A REACTION

There are a number of **factors which influence the rate of a reaction.** These are :

1. *Concentration of the reactants*
2. *Temperature*
3. *Nature of the reacting substances*
4. *Presence of catalyst*
5. *Exposure to radiations.*

In this chapter, we shall only study how the rate of a chemical reaction is influenced by *concentration of the reactants* and *temperature.*

It is observed that other factors remaining the same, the rate of a chemical reaction increases with the increase in concentration of the reactants. For example, we find that a piece of wood burns at a much faster rate in oxygen than in air. It is because of higher concentration of oxygen in the former.

The rate of reaction of almost all reactions increases with the increase in temperature. In most of the cases the rate of the reaction becomes almost double for every 10°C rise of temperature.

EXPERIMENT 3.1

AIM: ***To study the effect of concentration on the rate of reaction between sodium thiosulphate and hydrochloric acid.***

THEORY

According to the **law of mass action,** *rate of a chemical reaction is directly proportional to the product of the molar concentrations of the reactants.* In other words, the rate of reaction increases with the increase in the concentration of the reactants. The effect of concentration of reactants on rate of a reaction can be studied easily by the reaction between sodium thiosulphate and hydrochloric acid.

$$Na_2S_2O_3 + 2HCl \longrightarrow S(s) + 2NaCl(aq) + SO_2(g) + H_2O(l)$$

The insoluble sulphur, formed during the reaction, gives a milky appearance and makes the solution opaque. Therefore, rate of the reaction can be studied by measuring the time taken to produce enough sulphur to make some mark invisible on a paper kept under the conical flask in which the reaction is carried out.

APPARATUS

Stop-watch, two burettes and five conical flasks (100 ml).

MATERIALS REQUIRED

0.1 M $Na_2S_2O_3$ solution and 1 M HCl solution.

PROCEDURE

1. Wash the conical flasks with water and label them as 1, 2, 3, 4 and 5 respectively.
2. With the help of a burette, add 10, 20, 30, 40 and 50 ml of 0.1 M $Na_2S_2O_3$ solution to the flasks 1, 2, 3, 4 and 5 respectively.
3. Now add 40, 30, 20 and 10 ml of distilled water to the flask 1, 2, 3 and 4 respectively so that volume of solution in each flask is 50 ml.
4. Take 10 ml of 1 M HCl in a test tube with the help of a burette.
5. Add 10 ml of hydrochloric acid taken in a test tube to the conical flask No. 1 containing 10 ml of 0.1 M $Na_2S_2O_3$ and 40 ml of distilled water and start the stop-watch. When half of the hydrochloric acid solution has been added. Shake the contents of the conical flask and place it on the tile with a cross mark as shown in Fig. 3.1.
6. Go on observing from top to downwards in the flask and stop the stop-watch when the cross mark just becomes invisible. Note down the time.
7. Repeat the experiment by adding 10 ml of 1 M HCl to flasks 2, 3, 4 and 5 and record the time taken in each case for the cross to become just invisible.

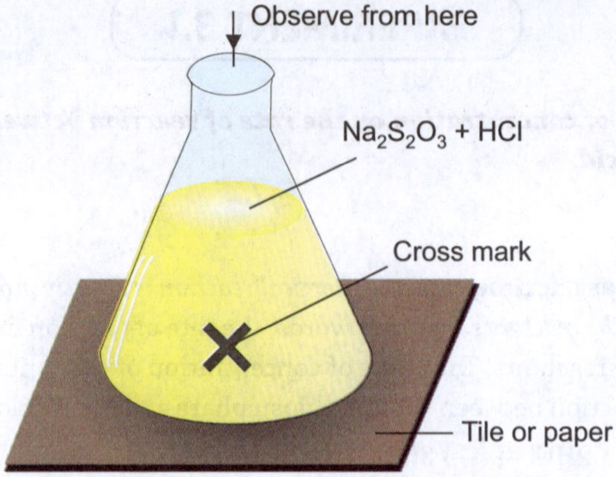

Fig. 3.1. Study of rate of reaction.

OBSERVATIONS

Flask No.	Volume of $Na_2S_2O_3$ solution in ml	Volume of water in ml	Total volume of solution in ml	Concentration of $Na_2S_2O_3$ solution	Volume of 1 M HCl (ml)	Time taken for cross to become just invisible (t)	$\frac{1}{t}$ (s^{-1})
1.	10	40	50	0.02 M	10	...... s	
2.	20	30	50	0.04 M	10	...... s	
3.	30	20	50	0.06 M	10	...... s	
4.	40	10	50	0.08 M	10	...... s	
5.	50	0	50	0.10 M	10	...... s	

PLOTTING OF GRAPH

Plot a graph between $\frac{1}{t}$ (in seconds) and the conc. of sodium thiosulphate by taking $\frac{1}{t}$ along ordinate (vertical axis) and conc. of $Na_2S_2O_3$ along abscissa (horizontal axis). It should be a straight sloping line.

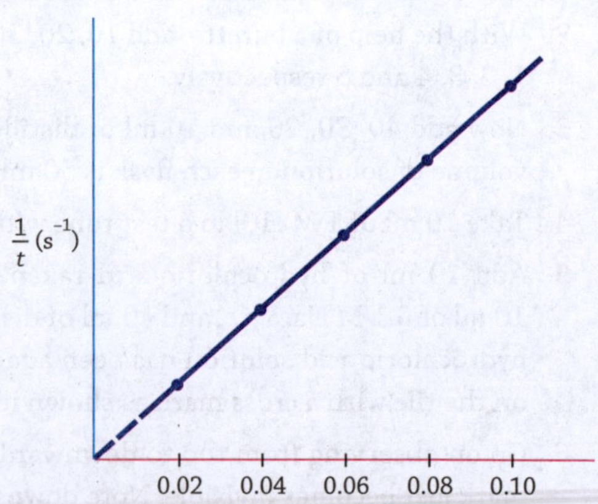

Fig. 3.2. A graph of conc. of $Na_2S_2O_3$ vs $\frac{1}{t}$.

RESULT

From the graph, it is clear that $\frac{1}{t}$ is directly proportional to the conc. of $Na_2S_2O_3$ solution. But $\frac{1}{t}$ is a direct measure of rate of the reaction, therefore, rate of the reaction between $Na_2S_2O_3$ and HCl is directly proportional to the conc. of $Na_2S_2O_3$ solution taken. Hence, rate of this reaction is directly proportional to the concentration of $Na_2S_2O_3$, which is one of the reactants.

Note: *It may be noted that the reaction rate also increases when the amount of sodium thiosulphate is kept constant but the concentration of hydrochloric acid is increased.*

PRECAUTIONS

1. The apparatus must be thoroughly clean. If the same conical flask is to be used again and again, it should be thoroughly washed with conc. HNO_3 and then with water.
2. Measure the volumes of sodium thiosulphate solution, hydrochloric acid and distilled water very accurately.
3. Use the same tile with the same cross-mark for all observation.
4. Complete the experiment at one time only so that there is not much temperature variation.
5. Start the stop-watch immediately when half of the hydrochloric acid solution has been added to sodium thiosulphate solution.
6. View the cross-mark through the reaction mixture from top to bottom from same height for all observations.

EXPERIMENT 3.2

AIM: ***To study the effect of change in temperature on the rate of reaction between sodium thiosulphate and hydrochloric acid.***

THEORY

The rate of a chemical reaction depends to a great extent upon temperature. The rate of reaction increases with increase in temperature. Increase in temperature increases kinetic energy of the molecules. Therefore, the fraction of molecules having energy greater than its threshold energy increases which results in the increase in number of effective collisions per second. It has been observed that in most of the cases *for every 10°C rise in temperature, the rate of the reaction becomes almost double.* The rate of reaction between sodium thiosulphate and hydrochloric acid also increases with increase in temperature.

APPARATUS

Conical flask (250 ml), measuring cylinders (50 ml and 5 ml), stop-watch, thermometer, tripod stand, wire-gauze and burner.

MATERIALS REQUIRED

0.1 M $Na_2S_2O_3$ solution, 1 M HCl, distilled water and conc. HNO_3.

PROCEDURE

1. Take 50 ml of 0.1 M $Na_2S_2O_3$ solution in a 100 ml conical flask and note its temperature with the help of a thermometer.
2. Add 10 ml of 1 M HCl to it and start the stop-watch immediately when half of the hydrochloric acid solution has been added.
3. Shake the contents of the flask gently and place it on the tile with a cross-mark as shown in Fig. 3.1.
4. Observe the cross-mark from the top and note the time taken for the mark to become just invisible.

5. Empty the flask and clean it thoroughly with conc. HNO_3 and then with water.
6. Take again 50 ml of 0.1 M $Na_2S_2O_3$ in conical flask and heat it so that the temperature of the solution becomes (T + 10°)C.
7. Remove the flask from the tripod-stand and add 10 ml of 1 M HCl to it and start the stop-watch.
8. Shake the contents gently and place it on the tile having a cross-mark.
9. Note the time taken for the mark to become just invisible.
10. Repeat the experiment at (T + 20)°C, (T + 30)°C and (T + 40)°C temperatures and record the observations as given below.

OBSERVATIONS

Volume of 0.1 M $Na_2S_2O_3$ solution taken each time = 50 ml

Volume of 1 M HCl added each time = 10 ml.

S.No.	Temperature	Time taken for cross to become just invisible (t)	$\frac{1}{t}$
1.			
2.			
3.			
4.			
5.			

PLOTTING OF GRAPH

Plot a graph by taking $\frac{1}{t}$ along the ordinate (vertical axis) and temperature along the abscissa (horizontal axis).

RESULT

Rate of reaction between sodium thiosulphate and hydrochloric acid increases with the increase in temperature.

PRECAUTIONS

Same as in experiment 3.1.

EXPERIMENT 3.3

AIM: ***To study the reaction rate of reaction of iodide ions with hydrogen peroxide at different concentrations of iodide ions.***

THEORY

Hydrogen peroxide oxidizes iodide ions to iodine in acidic medium

$$H_2O_2 + 2I^- + 2H^+ \longrightarrow 2H_2O + I_2$$

The reaction is monitored by adding a known volume of sodium thiosulphate solution and starch solution to the reaction mixture. Iodine liberated at once reacts with sodium thiosulphate solution and is reduced to iodide ions

$$I_2 + 2S_2O_3^{2-} \xrightarrow{\text{Fast}} S_4O_6^{2-} + 2I^-$$

When thiosulphate ions are completely consumed, the liberated iodine reacts with starch solution and gives blue colour

$$I_2 + \text{Starch} \longrightarrow \text{Blue complex}$$

The time elapsed before the appearance of blue colour, gives an idea about the rate of the reaction.

APPARATUS AND CHEMICALS

4 Conical flasks (250 ml), measuring cylinder, burette, pipette (5 ml), stop-watch, 0.1 M KI solution, 2.5 M H_2SO_4, starch solution, '3%' H_2O_2 solution, 0.05 M sodium thiosulphate solution.

PROCEDURE

1. Take four 250 ml conical flasks and label them as A, B, C and D.
2. Add 10 ml, 20 ml, 40 ml and 60 ml of 0.1 M KI solution to the flasks A, B, C and D respectively.
3. Add 10 ml of 2.5 M H_2SO_4 to each flask.
4. Add water to make the volume of solution 100 ml in each flask.
5. Add 5 ml starch solution to each flask.
6. Add 10 ml of 0.05 M sodium thiosulphate solution to each flask with the help of a burette.
7. Add 5 ml of 3% hydrogen peroxide solution to flask A with the help of a pipette and start the stop watch immediately. Stir the mixture and watch for the blue colour to appear. Note the time when the blue colour just appears.
8. Repeat the step 7 with the solutions in flasks B, C and D.
9. Record the observations in tabular form.

OBSERVATIONS

Flask	*0.1 M KI solution (ml)*	*2.5 M H_2SO_4 (ml)*	*Water (ml)*	*Starch solution (ml)*	*0.5 M sodium thiosulphate solution (ml)*	*3% hydrogen peroxide solution (ml)*
A	10	10	80	5	10	5
B	20	10	70	5	10	5
C	40	10	50	5	10	5
D	60	10	30	5	10	5

Time required for the blue colour to first appear in :

Flask A — .... *s*

Flask B — .... *s*

Flask C — .... *s*

Flask D — .... *s*

CONCLUSION

The rate of the reaction increases with increase in concentration of iodide ions.

PRECAUTIONS

1. Always use a freshly prepared solution of sodium thiosulphate.
2. Concentration of KI solution should be higher than the concentration of sodium thiosulphate solution.
3. Use freshly prepared starch solution.
4. Do not suck hydrogen peroxide solution with mouth but use a pipetter.

EXPERIMENT 3.4

AIM: ***To study the reaction rate of the reaction between potassium iodate (KIO_3) and sodium sulphite (Na_2SO_3) using starch solution as indicator.***

THEORY

In acidic medium, potassium iodate is reduced to iodide by sulphite. The reaction takes place through the following steps :

Step I : $IO_3^- + 2SO_3^{2-} \longrightarrow I^- + 2SO_4^{2-}$ ......(*Slow*)

Step II : $IO_3^- + 5I^- + 6H^+ \longrightarrow 3I_2 + 3H_2O$ ......(*Fast*)

Step III : $I_2 + SO_3^{2-} + H_2O \longrightarrow SO_4^{2-} + 2I^- + 2H^+$ ......(*Very fast*)

Sulphite ions react with potassium iodate producing iodide ions. Iodide ions, thus formed, are oxidized to iodine by reaction with more iodate ions. Iodine formed in Step II reacts immediately with sulphite ions forming iodide ions. When sulphite ions are completely consumed, the liberated iodine will not be consumed and would give blue colour, if strach is present. Thus, the above reaction can be monitored by adding a known but limited volume of sodium sulphite solution and starch solution. This is an example of **clock reaction** as the rate of the reaction is estimated by the time taken for the appearance of blue colour.

APPARATUS AND CHEMICALS

4 Conical flasks (250 ml), measuring cylinder, burette, pipette (25 ml), stop-watch.

0.01 M sodium sulphite solution, 0.1 M potassium iodate solution, starch solution, 2 M H_2SO_4.

PROCEDURE

1. Take four 250 ml conical flasks and label them as A, B, C and D.
2. Add 10 ml, 20 ml, 30 ml and 40 ml of 0.1 M KIO_3 solution to the flasks A, B, C and D respectively with the help of burette.
3. Add 10 ml of 2 M H_2SO_4 to each flask.
4. Add water to make the volume of solution 100 ml in each flask.
5. Add 5 ml of freshly prepared starch solution to each flask.

6. Add 25 ml of 0.01 M sodium sulphite solution to flask A with the help of a pipette and start the stop-watch immediately. Note the time when the blue colour just appears.
7. Repeat the step 6 with the solutions of flasks B, C and D.
8. Record the observations in tabular form.

OBSERVATIONS

Flask	0.1 M KIO_3 solution (ml)	2M H_2SO_4 (ml)	Water (ml)	Starch solution (ml)	0.01 M Na_2SO_3 solution (ml)	Time for appearance of blue colour
A	10	10	80	5	25	...... s
B	20	10	70	5	25	...... s
C	30	10	60	5	25	...... s
D	40	10	50	5	25	...... s

CONCLUSION

The rate of reaction increases with the increase in the concentration of potassium iodate.

PRECAUTIONS

1. Always use a freshly prepared solution of sodium sulphite because it is easily oxidized by air.
2. Concentration of KIO_3 solution should be higher than the concentration of sodium sulphite solution.
3. Use a freshly prepared starch solution.

VIVA QUESTIONS WITH ANSWERS

Q. 1. What is chemical kinetics ?

Ans. Chemical kinetics is the branch of chemistry which deals with the study of the rates of reactions and their mechanisms.

Q. 2. What do you understand by the rate of reaction ?

Ans. The rate of a reaction is defined as the change in the molar concentration of any one of the reactant or the product per unit time.

Q. 3. What are the units of the rate of reaction ?

Ans. mol L^{-1} s^{-1} (moles per litre per second).

Q. 4. What are the factors on which the rate of a reaction depends ?

Ans. The rate of reaction depends upon : *(i)* Nature of the reactants ; *(ii)* Concentration of the reactants ; *(iii)* Temperature ; *(iv)* Presence of catalyst ; and *(v)* Presence of radiations.

Q. 5. What is the law of mass action ?

Ans. Law of mass action states that the rate of a reaction is directly proportional to the product of molar concentrations of the reactants.

Q. 6. What is temperature coefficient of a reaction ?

Ans. Temperature coefficient of a reaction is the ratio of rate constants at two temperatures differing by 10°.

Q. 7. What are the units of rate constant for zero order reactions ?

Ans. Same as the rate of reaction, *i.e.* mol L^{-1} s^{-1}.

Q. 8. What are the units of rate constant for first order reactions ?

Ans. s^{-1}.

Q. 9. What is the effect of temperature on the rate constant of a reaction ?

Ans. It generally increases with increase in temperature.

Q. 10. Why certain reactions are very fast ?

Ans. Because they have very low activation energy.

Q. 11. What is threshold energy ?

Ans. It is the minimum energy which the colliding molecules must possess so as to have effective collision.

Q. 12. Why reactions with molecularity more than three are rare ?

Ans. Because simultaneous collision between more than three particles is rare on the basis of probability considerations.

Q. 13. "For an exothermic reaction activation energy for the forward reaction is less than that for the backward reaction." Is this statement true or false ?

Ans. True.

Q. 14. What is the rate determining step ?

Ans. In complex reactions, the slowest step determines the over all rate of the reaction. This step is known as the rate determining step.

Q. 15. What is the effect of catalyst on the activation energy and enthalpy change of the reaction ?

Ans. A catalyst decreases the activation energy of the reaction. It has no effect on the enthalpy change of the reaction.

Q. 16. On increasing the concentration of reactants the rate of the reaction does not change. What can you say about the order of the reaction from this observation ?

Ans. It is a zero order reaction.

Q. 17. Can order of a reaction be fractional ?

Ans. Yes. For example, for the reaction

$$CH_3CHO \longrightarrow CH_4 + CO$$

the order is equal to 3/2.

Q. 18. What is a complex reaction ?

Ans. A reaction involving more than one step is called a complex reaction.

Q. 19. What do you understand by '4 volume' H_2O_2 solution ?

Ans. 1 litre of '4 volume' H_2O_2 solution gives 4 litres of oxygen at N.T.P. on decomposition.

Q. 20. Express the conc. of 1 M H_2O_2 solution in terms of 'volume' strength.

Ans.

$$\underset{2\ \text{mol}}{2H_2O_2} \longrightarrow 2H_2O + \underset{22.4\ \text{L at N.T.P.}}{O_2}$$

1 litre of 1 M H_2O_2 contains 1 mole of H_2O_2 and at N.T.P. hence would give 11.2 L of oxygen at N.T.P., on complete decomposition. Hence, 1 M H_2O_2 solution is ***'11.2 volume'***.

Q. 21. What is the equivalent mass of H_2O_2 ?

Ans. 17.

Q. 22. What is the normality of 1 M H_2O_2 solution ?

Ans. 2 N.

Q. 23. What is the oxidation number of oxygen in H_2O_2 ?

Ans. – 1.

Q. 24. What is the colour of starch-iodine complex ?

Ans. Blue.

Q. 25. What is the effect of increase in conc. of iodide ions on the following reaction :

$$2H_3O^+ + 2I^- + H_2O_2 \longrightarrow 4H_2O + I_2$$

Ans. Rate of the reaction increases.

Q. 26. The reaction under examination is as follows :

$$S_2O_3^{2-}(aq) + 2H^+(aq) \longrightarrow H_2O(l) + SO_2(g) + S(s)$$

Write the conditions under which the rate law expression for this reaction can be written in the following manner.

Rate of precipitation of sulphur = $k[S_2O_3^{2-}]\ [H^+]^2$.

Ans. None of the reactants should be used in excess and the reaction should be elementary.

Q. 27. Suppose the above rate law expression for the precipitation of sulphur holds good, then on doubling the concentration of $S_2O_3^{2-}$ ion and H^+ ion, by how many times will the rate of the reaction increase ?

Ans. By eight times.

CHAPTER 4

Thermochemistry

4.01. ENTHALPY OF DISSOLUTION

It is well known that when a solute is dissolved in a solvent, heat is either *absorbed* or *evolved*. Thus, dissolution of a solute in a solvent is accompanied by enthalpy change (ΔH) of the system. If heat is *absorbed* (*i.e.*, the solution gets cooled), ΔH is given a *positive sign*. If heat is *evolved* (*i.e.*,the solution gets warmed), ΔH is given negative sign.

The enthalpy change per mole of a solute dissolved varies with the concentration of the solution. Therefore, it is necessary to express the enthalpy change with reference to the concentration of the solution. The **enthalpy of dissolution** is defined as *the enthalpy change per mole of a solute when it is dissolved in a pure solvent to give a solution of specified concentration*. For example, when one mole of anhydrous calcium chloride is dissolved in 400 moles of water, 78.60 kJ heat is *evolved*. A thermochemical equation to express this can be written as under :

$$CaCl_2(s) + 400H_2O(l) \longrightarrow CaCl_2(400\ H_2O)\ ;\ \Delta H = -\ 78.60\ kJ$$

Similarly, when one mole of hydrated calcium chloride ($CaCl_2.6H_2O$) is dissolved in 400 moles of water, 18.83 kJ heat is *absorbed*. The thermochemical equation to express this is written as :

$$CaCl_2.6H_2O(s) + 400\ H_2O(l) \longrightarrow CaCl_2.6H_2O\ (400\ H_2O)\ ;\ \Delta H = +\ 18.83\ kJ$$

The enthalpies of solution for some of the solutes dissolved in specified moles of water at 298 K are given in Table 4.1.

Table 4.1. Enthalpies of Solution in Water at 298 K

Substance (solute)	*Moles of H_2O (solvent)*	*Enthalpy of solution (kJ mol^{-1})*
NH_4Cl	200	+ 16.22
KCl	200	+ 18.58
NaCl	200	+ 5.35
KNO_3	200	+ 35.40
$CaCl_2$	400	– 78.60
$CaCl_2.6H_2O$	400	+ 18.83
$CuSO_4.5H_2O$	400	+ 11.5
$CuSO_4$	400	– 66.0

4.02. ENTHALPY OF NEUTRALISATION

Enthalpy of neutralisation of an acid at a given temperature is defined as *enthalpy change (ΔH) accompanying the neutralisation of one gram equivalent of the acid by a base in dilute solutions at* that temperature. Enthalpy of neutralisation of an acid may also be defined as *the enthalpy change accompanying the formation of one mole of water by reaction between the acid and a base in dilute solutions.* The neutralisation of hydrochloric acid by sodium hydroxide in dilute solutions at 298 K is represented by the thermochemical equation.

$$HCl(aq) + NaOH(aq) \longrightarrow NaCl(aq) + H_2O(l)\ ;\ \Delta H = -57.32\ kJ$$

Thus, the enthalpy of neutralisation of hydrochloric acid by sodium hydroxide at 298 K is – 57.32 kJ.

Similarly, *enthalpy change accompanying neutralisation of one gram equivalent of a base by an acid in dilute solutions at a given temperature is known as the* **enthalpy of neutralisation of the base** at that temperature. In the above example, the enthalpy of neutralisation of sodium hydroxide with hydrochloric acid is also – 57.32 kJ at 298 K.

The neutralisation of hydrochloric acid by sodium hydroxide in dilute solutions, when the acid, alkali and the salt formed are completely dissociated, may be represented as

$$H^+(aq) + Cl^-(aq) + Na^+(aq) + OH^-(aq) \longrightarrow Na^+(aq) + Cl^-(aq) + H_2O(l)$$

or

$$H^+(aq) + OH^-(aq) \longrightarrow H_2O(l)$$

Neutralisation of strong acid and strong base involves the combination of H^+ and OH^- ions to form unionised water. It is, therefore, expected that the enthalpy of neutralisation of every strong acid by a strong base should be same and this is largely so, as is evident from the data in Table 4.2.

Table 4.2. Enthalpies of Neutralisation of Strong Acids by Strong Bases at 298 K

Acid	*Alkali*	*Enthalpy of Neutralisation (ΔH)*
HCl	NaOH	– 57.32 kJ
HCl	KOH	– 57.45 kJ
HNO_3	NaOH	– 57.28 kJ

4.03. CALORIMETER CONSTANT OF CALORIMETER

Measurement of heat changes are carried out in calorimeters. During measurement of heat changes, calorimeter also absorbs some heat which is expressed in terms of calorimeter constant. **Calorimeter constant** is defined as *the amount of heat required to raise the temperature of the calorimeter by one degree Celsius.* It is denoted by W and has units joules per degree Celsius or joules per kelvin.

EXPERIMENT 4.1

AIM: ***Determine the calorimeter constant (W) of calorimeter (polythene bottle).***

THEORY

In order to determine the calorimeter constant, a known volume of hot water at a known temperature is added to a known volume of water taken in the calorimeter at room temperature. Since energy is conserved, the heat gained by the calorimeter and the cold water must be equal to the heat lost by hot water. If t_1, t_2 and t_3 are the temperatures of cold water, hot water and mixture respectively and m_1, m_2 and m_3 are the masses of calorimeter, cold water and hot water respectively, then we can write

$$m_1C_1(t_3 - t_1) + m_2C(t_3 - t_1) = m_3C(t_2 - t_3)$$

Here C_1 is the heat capacity of calorimeter and C is the heat capacity of water. The quantity m_1C_1 is the calorimeter constant, W

$$W(t_3 - t_1) + m_2C(t_3 - t_1) = m_3C(t_2 - t_3)$$

$$W(t_3 - t_1) = m_3C(t_2 - t_3) - m_2C(t_3 - t_1)$$

$$W = \frac{C[m_3(t_2 - t_3) - m_2(t_3 - t_1)]}{(t_3 - t_1)}$$

The heat capacity of water, C is 4.184 J/K or J/°C.

Knowing all the parameters on RHS, the value of W can be calculated.

It may be noted that $t_2 > t_3 > t_1$.

PROCEDURE

1. Put 100 ml of distilled water in polythene bottle with a thermometer and stirrer Fig. 4.1.

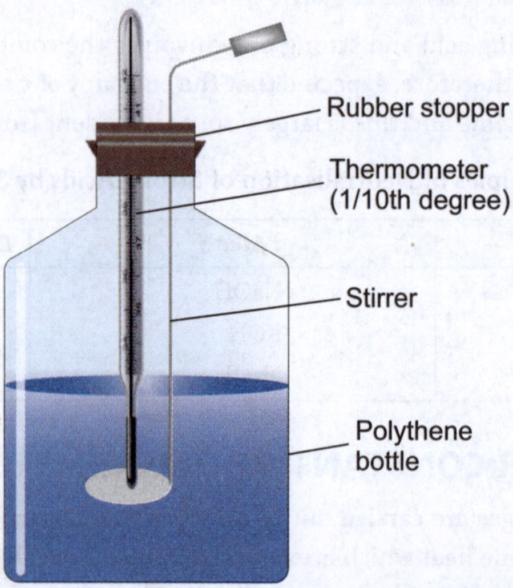

Fig. 4.1. Polythene bottle calorimeter.

2. Note the temperature (t_1°C).
3. Heat some water in a beaker to a temperature 20–30°C higher than that of room temperature.
4. Put 100 ml of this warm water in another beaker.
5. Note the temperature of this water. Let it be t_2°C.
6. Add warm water from the beaker into the polythene bottle without any loss of time.
7. Stir the contents.
8. Read the temperature attained after mixing. Let it be t_3°C.

OBSERVATIONS

Volume of water taken in bottle	= 100 ml
Temperature of water	= t_1°C

Volume of warm water added = 100 ml

Temperature of warm water = t_2°C

Temperature after mixing = t_3°C

CALCULATIONS

Heat given out by hot water = Heat taken by bottle and cold water.

$$-100 \times 4.184 \times (t_3 - t_2) = W \times (t_3 - t_1) + 100 \times 4.184 \times (t_3 - t_1)$$

$$W = \frac{100 \times 4.184 \times (t_2 - t_3) - 100 \times 4.184(t_3 - t_1)}{(t_3 - t_1)}$$

$$W = 4.184 \left[100 \frac{(t_2 - t_3)}{(t_3 - t_1)} - 100 \right] \text{J/°C}$$

From the above expression the calorimeter constant, W can be calculated.

EXPERIMENT 4.2

AIM: ***Determine the enthalpy of dissolution of given solid copper sulphate ($CuSO_4.5H_2O$) in water at room temperature.***

THEORY

In this experiment, the enthalpy of dissolution is measured by the use of calorimetric techniques. A known volume of the water is taken in a polythene bottle as shown in Fig. 4.1. Its temperature is noted and then known weight of the solute is added to it. The solution is stirred gently and change in temperature is recorded. From the change in temperature, heat absorbed or evolved can be calculated. In this experiment one mole of solute is dissolved per 400 moles of water. For maintaining this ratio 7.0 g of $CuSO_4.5H_2O$ is dissolved in 200 ml of water.

REQUIREMENTS

(*a*) **Apparatus.** 250 ml or 500 ml polythene bottle fitted with a rubber cork with two holes, one for thermometer (1/10th degree) and other for stirrer, two beakers, stirrer and measuring cylinder.

(*b*) **Chemicals.** Hydrated copper sulphate, distilled water.

PROCEDURE

A. Determination of Calorimeter Constant

Calorimeter constant is determined by following the procedure as described in Experiment 4.1.

B. Determination of Enthalpy of Dissolution

1. Put 200 ml of distilled water into the polythene bottle.
2. Now fit a cork with two holes into the mouth of the polythene bottle. Insert a thermometer into one hole with its bulb about 1 cm above the bottom of the bottle. Put the stirrer into the second hole.

3. Note down the temperature (t_1).
4. Take a known weight of finely powdered substance.
5. Transfer the known weight (say w g) of finely powdered hydrated copper sulphate quickly by removing the rubber cork and putting it back into its position without any loss of time.
6. Stir it with the help of a stirrer till hydrated copper sulphate is dissolved. However, *the rate of stirring should be kept as low as efficiency permits to minimise the energy introduced by stirring* (vigorous stirring does cause some increase in temperature).
7. Note down the temperature (t_2) when the substance just dissolves.

OBSERVATIONS

Weight of the hydrated copper sulphate dissolved = w g

Volume of water taken into the bottle = 200 ml

= 200 g (assuming density = 1 g/ml)

Temperature of water = t_1°C

Temperature of water after dissolving hydrated copper sulphate = t_2°C

Calorimeter constant of the polythene bottle = W J/°C

CALCULATIONS

Assuming density and specific heat of the solution to be same as that of water, heat evolved or absorbed for dissolution of w g of the solute

$$Q = W(t_2 - t_1) + (200 + w)\,(t_2 - t_1) \times 4.184 \text{ J}$$

Heat liberated on dissolution of 1 g of copper sulphate

$$= \frac{W(t_2 - t_1) + (200 + w)\,(t_2 - t_1) \times 4.184 \text{ J}}{w}$$

Heat liberated on dissolution of 1 mol (249.5 g) of copper sulphate

$$= \frac{W(t_2 - t_1) + (200 + w)\,(t_2 - t_1) \times 4.184}{w} \times 249.5 \text{ J}$$

$$\therefore \Delta_{sol} H \text{ of copper sulphate} = -\frac{W(t_2 - t_1) + (200 + w)\,(t_2 - t_1) \times 4.184}{w} \times 249.5 \text{ J/mol}$$

RESULT

Enthalpy of dissolution of copper sulphate is ...... J/mol.

Note: If $t_2 > t_1$, heat is evolved during dissolution and $D_{sol}H$ has negative sign.

Similarly we can find out the enthalpy of dissolution of potassium nitrate. For that dissolve 5.5 g of KNO_3 in 200 mL of water. Here, the mole ratio of solute and solvent is 1 : 200.

EXPERIMENT 4.3

AIM: ***Determine the enthalpy of neutralisation of hydrochloric acid with sodium hydroxide solution.***

THEORY

Heat is evolved during neutralisation of an acid with an alkali. Known volumes of the standard solutions of an acid and alkali are mixed and the change in temperature is observed and from this, the enthalpy of neutralisation is calculated. *Enthalpy of neutralisation is the heat evolved when one gram equivalent of the acid is completely neutralised by a base in dilute solution.*

REQUIREMENTS

(*a*) **Apparatus.** A wide-mouthed polythene bottle (to serve as calorimeter), a rubber cork having two holes, thermometer (1/10th degree), stirrer fitted with a cork on the handle, and a 100 ml measuring cylinder.

(*b*) **Chemicals.** 1.0 M hydrochloric acid and 1.0 M sodium hydroxide solution.

PROCEDURE

A. Determination of Calorimeter Constant

Calorimeter constant is determined by following the procedure as described in Experiment 4.1.

B. Determination of Enthalpy of Neutralisation

1. Clean and dry the polythene bottle.
2. Place 100 ml of 1.0 M hydrochloric acid solution in it.
3. Record the temperature of the acid solution.
4. Similarly, note the initial temperature of the sodium hydroxide solution taken in a separate vessel.
5. Both the solutions should have the same temperature, otherwise wait for some time so that they attain the same temperature.
6. Transfer 100 ml of sodium hydroxide solution into the acid solution quickly.
7. Immediately fit the cork having the thermometer and the stirrer in the mouth of the polythene bottle (Fig. 4.1) and stir well.
8. Note the temperature after small intervals till it becomes constant.
9. Record the highest temperature (to 0.1°) reached.

OBSERVATIONS

Initial temperature of the acid and base	$= t_1$°C
Final temperature after neutralisation	$= t_2$°C
Change in temperature, Δt	$= (t_2 - t_1)$°C
Mass of the mixture solution after neutralisation	= 200 g*
Calorimeter constant of calorimeter	= W J/°C

*It is assumed that the solution has same density as water.

CALCULATIONS

Heat produced during neutralisation of 100 ml of 1.0 M HCl

$$= (200 + W) \times (t_2 - t_1) \times 4.184 \text{ joules}$$

∴ Heat produced during neutralisation of 1000 ml of 1 M HCl

$$= (200 + W) \times (t_2 - t_1) \times 4.184 \times \frac{1000}{100} \text{ joules}$$

$$= \frac{(200 + W) \times (t_2 - t_1) \times 4.184}{100} \text{ kJ}$$

Since heat is produced during neutralisation, the enthalpy of neutralisation is negative.

$$\therefore \quad \textbf{Enthalpy of neutralisation} = \frac{\mathbf{(200 + W) \times (t_2 - t_1) \times 4.184}}{\mathbf{100}} \textbf{ kJ.}$$

RESULT

The enthalpy of neutralisation of HCl with NaOH is ...... kJ.

Percentage error = ...

Note: *Enthalpy of neutralisation of all strong acids with strong bases and vice versa is – 57.3 kJ.*

It may be noted that 1000 mL of 1 M HCl contains 1 mole (or 1 equivalent) of HCl.

EXPERIMENT 4.4

AIM: ***Determine the enthalpy change during the interaction (hydrogen bond formation) between acetone and chloroform.***

THEORY

When acetone is mixed with chloroform, heat is evolved due to formation of hydrogen bonds between chloroform and acetone :

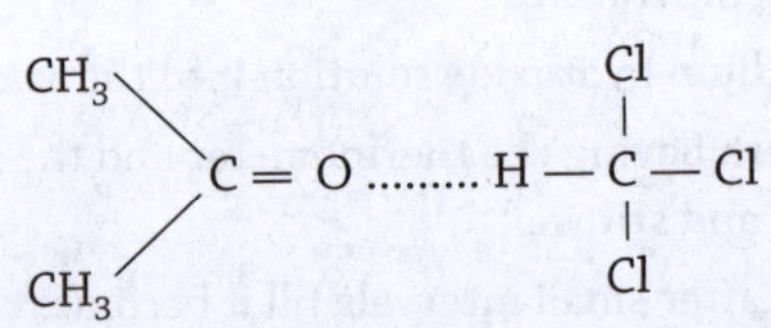

Heat evolved during this interaction can be determined experimentally by mixing the two liquids and measuring the heat change by using a calorimeter.

REQUIREMENTS

(a) Apparatus. A wide mouthed polythene bottle fitted with a thermometer (1/10th degree) and a stirrer (to serve as calorimeter), 100 ml measuring cylinder.

(b) Chemicals. Pure acetone and pure chloroform.

PROCEDURE

A. Determination of Calorimeter Constant

Calorimeter constant is determined as described in Experiment 4.1.

B. Determination of Enthalpy of Interaction of Acetone and Chloroform

1. Take a clean and dry polythene bottle calorimeter.
2. Place 100 ml acetone in it.
3. Note the temperature of acetone.
4. Take 100 ml of chloroform in a beaker and note its temperature. Both the solutions should have same temperature otherwise wait for sometime so that they attain same temperature.
5. Transfer the chloroform into the calorimeter and immediately fit the cork (or lid) having thermometer and stirrer. Stir gently.
6. Note the temperature after small intervals till it becomes constant.
7. Record the highest temperature reached.

OBSERVATIONS AND CALCULATIONS

Initial temperature of acetone and chloroform	$= t_1$°C
Final temperature after mixing the two liquids	$= t_2$°C
Change in temperature	$= (t_2 - t_1)$°C
Calorimeter constant of calorimeter	= W J/°C
Density of chloroform	= 1.499 g/cm^3
Density of acetone	= 0.787 g/cm^3
Heat capacity of chloroform, S_1	= 0.96 J/g
Heat capacity of acetone, S_2	= 2.18 J/g

Heat change = W × 4.184 × $(t_2 - t_1)$ + [100 × 1.499 × S_1 + 100 × 0.787 × S_2] $(t_2 - t_1)$ joules

= X joules

Since $t_2 > t_1$ in this experiment, heat is evolved and enthalpy change for the interaction of acetone and chloroform has negative sign.

RESULT

Enthalpy change during mixing of 100 ml of acetone with 100 ml of chloroform

= – X joules.

VIVA QUESTIONS WITH ANSWERS

Q. 1. Define enthalpy of neutralization.

Ans. Enthalpy of neutralization of an acid or a base is the enthalpy change when one gram equivalent of the acid is neutralised by a base or vice versa.

Q. 2. What is the value of enthalpy of neutralisation of a strong acid and a strong base ?

Ans. When one gram equivalent of a strong acid is neutralised by one mole of a strong base or *vice-versa*, the enthalpy change is always equal to – 57.3 kJ.

Q. 3. Why is the enthalpy of neutralisation of a strong acid with a strong base always the same ?

Ans. This is because it always involves the combination of one mole of H^+ ions with OH^- ions to form unionised water molecules.

Q. 4. Define enthalpy of solution.

Ans. It is the enthalpy change taking place when one mole of a substance is dissolved in a specified number of moles of solvent at a given temperature and pressure.

Q. 5. Why is copper sulphate taken in powdered form ?

Ans. To facilitate its dissolution in minimum time and thus preventing heat loss to the surroundings.

Q. 6. Will enthalpy of solution of hydrated copper sulphate and anhydrous copper sulphate be same ?

Ans. No, in case of anhydrous copper sulphate enthalpy change will not only correspond to the dissolution process but also to hydration process, *i.e.*, we get enthalpy of hydration plus enthalpy of solution.

Q. 7. Why is temperature recorded with a thermometer calibrated to 1/10th degree ?

Ans. For more accurate results.

Q. 8. Is the enthalpy of neutralisation of acetic acid the same as that of HCl ? If not why ?

Ans. Acetic acid is a weak acid and is not completely ionised. Some heat is used up for the ionisation of acetic acid. Hence, the net heat evolved is less and not the same as that of HCl which is completely ionised.

Q. 9. Is the dissolution of hydrated copper sulphate an exothermic or an endothermic process ?

Ans. Endothermic process.

Q. 10. 50 ml of a liquid A are mixed with 50 ml of liquid B. The volume of resulting solution is found to be 99.5 ml. What do you conclude about nature of the solution ?

Ans. The solution shows a negative deviation from Raoult's law, A—B interactions are stronger than A—A and B—B interactions.

Q. 11. When a liquid A is mixed with liquid B, the resulting solution is found to be cooler. What do you conclude about nature of the solution ?

Ans. The solution shows a positive deviation. Absorption of heat takes place. A—B interactions are weaker than A—A and B—B interactions.

Q. 12. What type of deviation is expected of a solution obtained by adding conc. H_2SO_4 to water ?

Ans. The solution shows negative deviation. Heat is liberated. A—B interactions are stronger than A—A and B—B interactions.

CHAPTER 5

Electrochemistry

When a redox reaction is carried out indirectly, the chemical energy is converted into electrical energy. *A device in which chemical energy is converted into electrical energy is called an* **electrochemical cell.** Electrochemical cells are also known as **galvanic cells** or **voltaic cells.** Daniell cell is an example of voltaic cell. In order to understand the basic principle involved, it is desirable to recall definitions of oxidation and reduction and to consider some oxidation reduction reactions taking place in beakers as well as in electrochemical cells.

OXIDATION

It is defined as *a process which involves loss of electrons by a substance.*

The substance which loses electrons is said to be *oxidised.*

REDUCTION

It is defined as *a process which involves gain of electrons by a substance.*

The substance which gains electrons is said to be *reduced.*

A substance cannot lose electrons unless another substance which can gain electrons is also present in its environment. In other words, oxidation can takes place only if reduction can also take place at the same time. This can be illustrated by taking examples below.

5.01. Zn-$CuSO_4$ REDOX REACTION IN A BEAKER (Oxidation of Zn metal by Cu^{2+} ions)

When a zinc rod is placed in a solution of copper sulphate as shown in Fig. 5.1 it is observed that :

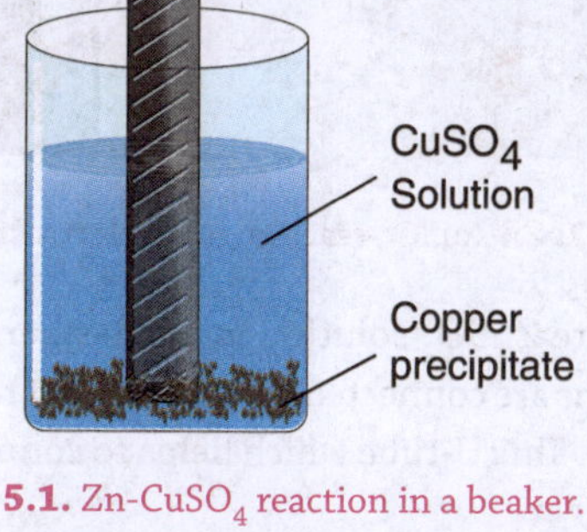

Fig. 5.1. Zn-$CuSO_4$ reaction in a beaker.

The zinc strip starts dissolving forming Zn^{2+} ions in solution. This indicates that the oxidation of zinc metal to zinc ions takes place, which is represented as :

$$Zn(s) \longrightarrow Zn^{2+}(aq) + 2e^- \quad ...(5.1)$$

Simultaneously, copper starts precipitating out from the solution, indicating the reduction of Cu^{2+} ions which are present in solution to copper metal. This may be represented as :

$$Cu^{2+}(aq) + 2e^- \longrightarrow Cu(s) \quad ...(5.2)$$

The reactions (5.1) and (5.2) are known as **half-reactions,** *i.e.*, oxidation half-reaction and reduction half-reaction.

The overall **redox reaction** is obtained by adding the two half-reactions (5.1) and (5.2) and is written as :

$$Zn(s) + Cu^{2+}(aq) \longrightarrow Zn^{2+}(aq) + Cu(s).$$

The reaction is accompanied by evolution of heat, *i.e.*, reaction is exothermic and thus ΔH is negative.

But if we try to immerse copper rod in a solution of zinc sulphate we shall observe hardly any change, concluding thereby that the reverse reaction is not possible. So *zinc metal can be oxidised by Cu^{2+} ions but copper metal cannot be oxidised by Zn^{2+} ions.*

5.02. REDOX REACTION IN AN ELECTROCHEMICAL CELL

In an electrochemical cell, the redox reaction takes place indirectly. Here, the oxidation and reduction take place in different vessels and the electrons are transferred from the reducing agent (the substance oxidized) to the oxidizing agent (the substance reduced) through connecting wires. Hence in such cells, chemical energy is converted into electrical energy. A simple laboratory form of electrochemical cell involving Zn-$CuSO_4$ reaction is shown in Fig. 5.2.

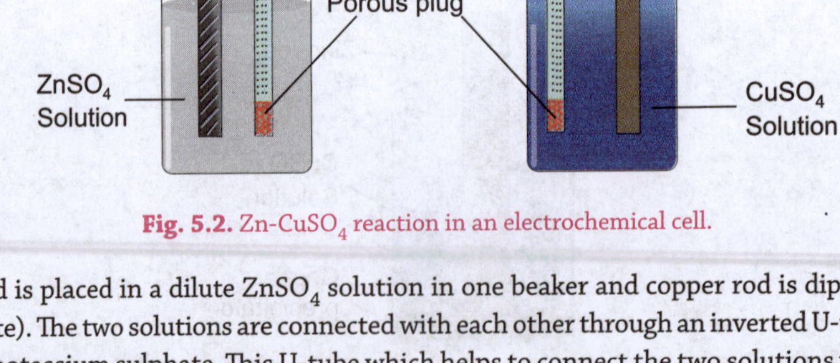

Fig. 5.2. Zn-$CuSO_4$ reaction in an electrochemical cell.

Zinc rod is placed in a dilute $ZnSO_4$ solution in one beaker and copper rod is dipped in a $CuSO_4$ solution (dilute). The two solutions are connected with each other through an inverted U-tube containing a solution of potassium sulphate. This U-tube which helps to connect the two solutions with each other is called **salt bridge.**

If the metal rods are not connected to each other or if the salt bridge is taken out, no current flows through the ammeter and no reaction takes place in the cell. But as soon as the connection is made as shown (Fig. 5.2), the current starts flowing as indicated by the ammeter reading and the chemical reaction takes place as :

$$Zn(s) + Cu^{2+}(aq) \longrightarrow Zn^{2+}(aq) + Cu(s)$$

The current continues to flow as long as the chemical reaction continues to take place.

Electricity cannot flow from one point to another unless there is a potential difference between the two points. Hence, the flow of electricity from one electrode to another electrode in any cell indicates that the two electrodes have different potentials. *The difference of potential which causes flow of current from one electrode (which is at a higher potential) to another electrode (which is at a lower potential) is called the* **electromotive force (EMF).** The EMF of a cell can be measured by connecting the two electrodes to the two terminals of a voltmeter. The potential difference is then read directly from the instrument. The EMF is also known as *cell potential* and is measured in units of volts.

EXPERIMENT 5.1

AIM: ***To set up simple Daniell cell and determine its EMF.***

THEORY

When a copper electrode dipped in copper sulphate solution is connected to a zinc electrode dipped in the zinc sulphate solution, then electrons flow from zinc electrode to copper electrode and the chemical reactions take place as :

$$Zn(s) \longrightarrow Zn^{2+}(aq) + 2e^-$$

$$Cu^{2+}(aq) + 2e^- \longrightarrow Cu(s)$$

$$\textit{Overall reaction}: Zn(s) + Cu^{2+}(aq) \longrightarrow Zn^{2+}(aq) + Cu(s)$$

APPARATUS AND CHEMICALS

One beaker, a porous pot, connecting wires, milli voltmeter, sand paper, zinc strip, copper strip, 1 M $ZnSO_4$ solution and 1 M $CuSO_4$ solution.

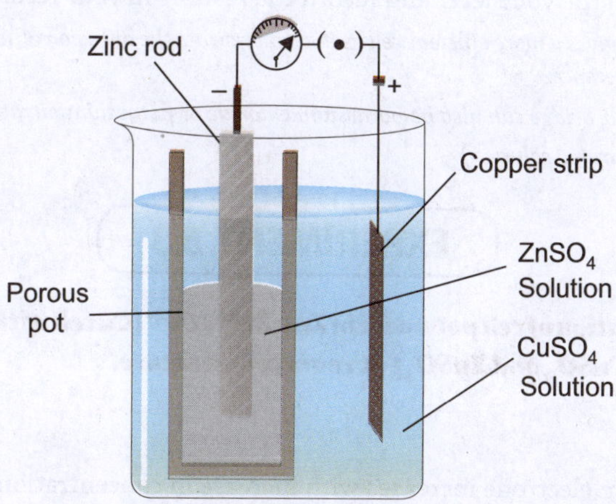

Fig. 5.3. A Daniell cell.

PROCEDURE

1. Take copper sulphate solution in a clean beaker.
2. Clean the copper strip with the help of sand paper and dip it into copper sulphate solution.
3. Take zinc sulphate solution in a porous pot.
4. Clean the zinc strip with the help of sand paper and dip it into zinc sulphate solution.
5. Keep the porous pot in the beaker.
6. Connect the copper strip with the positive terminal and zinc strip with the negative terminal of a voltmeter as shown in Fig. 5.3.
7. Note the position of the pointer in the voltmeter and record the reading in your note-book.

OBSERVATION

The EMF of the Daniell cell is ...... volts.

PRECAUTIONS

1. The concentration of copper sulphate and zinc sulphate should neither be too low nor too high.
2. The porous pot should not be completely dipped into the copper sulphate solution, *i.e.*, the copper sulphate solution should not be allowed to enter into the porous pot.
3. Clean zinc and copper strips with sand paper before use.
4. Carry out dilution of the solution carefully.
5. Note the reading only when the pointer becomes stable.
6. Connect copper strip with the positive terminal of voltmeter and zinc strip with negative terminal.

EXPERIMENT 5.2

AIM: ***To set up simple Daniell cell using salt bridge and determine its EMF.***

PROCEDURE

Set up the apparatus as shown in Fig. 5.2 containing 1 M solutions of $ZnSO_4$ and $CuSO_4$ and note the position of the pointer in the voltmeter and record the reading in your record book.

Notes: 1. *Use of salt bridge gives a more efficient cell as it even prevents the diffusion of solvent molecules resulting from concentration difference.*

2. *Electrolytes in salt bridge can also be potassium chloride or potassium nitrate containing agar-agar (3 g in 100 ml of the saturated solution).*

EXPERIMENT 5.3

AIM: ***To study the variation of cell potential in Zn| Zn^{2+} || Cu^{2+} |Cu cell with change in concentration of electrolytes ($CuSO_4$ and $ZnSO_4$) at room temperature.***

THEORY

Reduction potential of an electrode increases with increase in concentration of the electrolyte.

$$M^{n+}(aq) + ne^- \longrightarrow M(s)$$

In the zinc-copper electrochemical cell zinc electrode acts as anode while copper electrode acts as cathode.

$$E_{cell} = E_{cathode} - E_{anode}$$

E_{cell} increases if $E_{cathode}$ increases and E_{anode} decreases. Thus, using higher concentration of Cu^{2+} and lower conc. of Zn^{2+} ions increase the E_{cell} of $Zn \mid Zn^{2+} \mid\mid Cu^{2+} \mid Cu$.

The relation between concentration of the electrolyte and the standard electrode potential is given in the form of Nernst equation :

$$E = E^\circ - \frac{0.059}{n} \log \frac{[M]}{[M^{n+}]}$$

APPARATUS AND CHEMICALS

Same as in Experiment 5.1.

PROCEDURE

Same as in Experiment 5.1.

Repeat the experiment by taking different concentrations of zinc sulphate and copper sulphate solutions.

OBSERVATIONS

Concentration of $ZnSO_4$ solution	*Concentration of $CuSO_4$ solution*	*EMF of the cell*
1 M	1 M	 V
1 M	0.5 M	 V
1 M	0.025 M	 V
1 M	0.0125 M	 V
0.5 M	1 M	 V
0.025 M	1 M	 V
0.0125 M	1 M	 V

CONCLUSION

EMF of the cell increases with decrease in concentration of the electrolyte around anode and increase in concentration of the electrolyte around cathode.

VIVA QUESTIONS WITH ANSWERS

Q. 1. What is oxidation ?

Ans. It is a process which involves loss of electrons by a substance.

Q. 2. What is reduction ?

Ans. It is a process which involves gain of electrons by a substance.

Q. 3. What is a voltaic cell or electrochemical cell ?

Ans. It is a device of converting the chemical energy into electrical energy.

Q. 4. Define the term EMF ?

Ans. The difference of potential which causes flow of current from one electrode to another electrode in an electrochemical cell is called the electromotive force (EMF).

Q. 5. What is a half cell ?

Ans. It is half of the electrochemical cell where either oxidation or reduction occurs.

Q. 6. Mention the names of anode and cathode of a Daniell cell ?

Ans. Zinc strip acts as anode whereas copper strip as cathode.

Q. 7. Can copper be oxidised by zinc ions ?

Ans. No, copper metal cannot be oxidised by Zn^{2+} ions.

Q. 8. What is a salt bridge ?

Ans. A salt bridge is a device to keep internal continuity between the two half cells of a voltaic cell and to prevent the physical contact between the two electrolytes. It also maintains the electrical neutrality of the electrochemical cell.

Q. 9. Name the electrolytes that can be used in salt bridge.

Ans. Potassium sulphate, potassium chloride or potassium nitrate containing agar-agar.

Q. 10. What is the function of porous pot in a Daniell cell ?

Ans. The porous pot maintains the ionic continuity as well as prevents mixing of the two solutions.

Q. 11. What is the direction of flow of electrons in an electrochemical cell ?

Ans. Electrons move from anode to cathode in an electrochemical cell.

Q. 12. What is the direction of flow of current in an electrochemical cell ?

Ans. Current flows from cathode to anode.

Q. 13. What is the effect of increase in $[Zn^{2+}]$ on EMF of the cell $Zn \mid Zn^{2+} \parallel Cu^{2+} \mid Cu$?

Ans. The EMF decreases with the increase in molar concentration of Zn^{2+} ions.

Q. 14. What is the effect of increase in $[Cu^{2+}]$ on EMF of the cell $Zn \mid Zn^{2+} \parallel Cu^{2+} \mid Cu$?

Ans. The EMF increases with the increase in molar concentration of Cu^{2+} ions.

Q. 15. What is the sign of ΔG for the reaction in electrochemical cell ?

Ans. ΔG is – ve for the reaction in electrochemical cell.

Q. 16. What factor is kept in mind while selecting an electrolytic solution for the construction of a salt bridge ?

Ans. The ions of the electrolyte in the salt should not react with ions of the electrolytes around electrodes.

Q. 17. Is it possible to measure the single electrode potential ?

Ans. No, because reaction in a half cell does not take place independently.

Q. 18. What is a primary battery ?

Ans. A battery in which the reaction occurs only once and after use over a period of time becomes dead and cannot be used again. For example, dry cell.

Q. 19. What is secondary battery ?

Ans. A battery which can be recharged by passing current through it in the opposite direction and can be reused. For example, lead storage battery.

CHAPTER 6

Chromatography

Chromatography is a modern and sensitive technique used for rapid and efficient analysis and (or) separation of components of a mixture and purification of compounds.

The basic principle of chromatographic technique is based on the differential migration of the individual components of a mixture through a stationary phase under the influence of moving phase. The *stationary phase* may be a porous solid (such as silica, alumina, etc.) packed in a column or supported on a filter paper or a glass strip. The *moving* (or *mobile*) *phase* may be some solvent or a gas and is referred as an **eluent.**

6.01. TYPES OF CHROMATOGRAPHY

There are various types of chromatographic techniques which differ from one another on the basis of difference in the moving phase and the stationary phase.

1. Column Chromatography

(*a*) **Adsorption chromatography.** In this type, the mixture is dissolved in some suitable solvent such as alcohol, ether, benzene, etc. and the resulting solution is poured down a vertical column filled with the adsorbing material such as *alumina, chalk, charcoal, silica gel,* etc. The process of addition of the mixture to the column is called **loading.** Depending upon the rate at which different components of the mixture are adsorbed, different zones or bands are formed down the column ; substance which is adsorbed most, remains at the top of the column. Later, the constituents are washed down and collected separately, with a suitable solvent. This process is called **'elution'**.

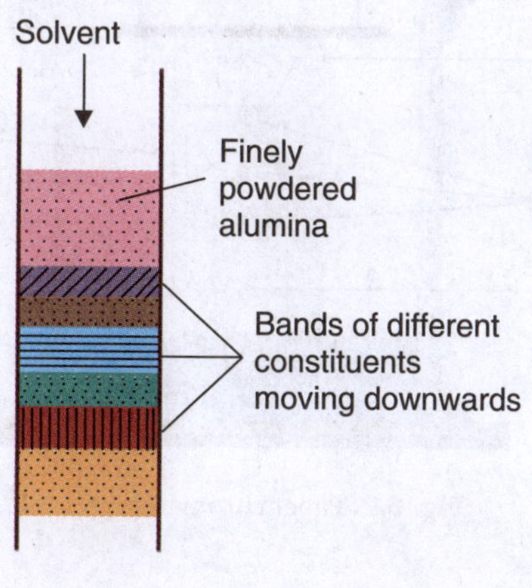

Fig. 6.1. Column chromatography.

(*b*) **Partition chromatography.** This is based on the principle that if to a mixture of two immiscible liquids A and B, a substance which is soluble in both A and B is added, then it distributes itself in such a way that the ratio of its concentrations in two liquids A and B is constant at a particular temperature. The technique used here is the same as for the adsorption type. Separation of different constituents takes place because each constituent distributes itself to different degrees between the solvent which flows down the column and the stationary liquid.

2. Thin Layer Chromatography (TLC)

In this technique the adsorbent (alumina or silica gel) is pasted on a thin strip of glass and is dried. The substance under investigation is dissolved in some suitable solvent A drop of this solution is put on one end of the glass plate and it is kept vertically in a vessel containing a solvent. Due to capillary action the solvent rises up carrying along with it the constituents which are adsorbed on the plate at different distances depending upon the extents of adsorption. The weakly adsorbed components rise to greater heights.

3. Gas Chromatography

This is relatively new technique used to analyse mixtures of gases, liquids and volatile solids. A small quantity of the mixture is introduced into a stream of a gas which acts as moving phase. The stream of gas along with the substance to be analysed passes through a column and different constituents come out one by one at different intervals and are recorded automatically on a chromatogram.

4. Paper Chromatography

It is mainly a type of partition chromatography in which a special adsorbent paper is used instead of a column. Moisture *adsorbed by this acts as a* **stationary phase** *and the solvent as a* **moving phase.** The mixture to be separated or analysed is put at one end of the paper strip as a small spot. The paper is placed in a container, with a suitable solvent, vertically in such a way that the lower end (where the mixture spot is put) dips in the solvent and the spot remains slightly above the solvent level (Fig. 6.2).

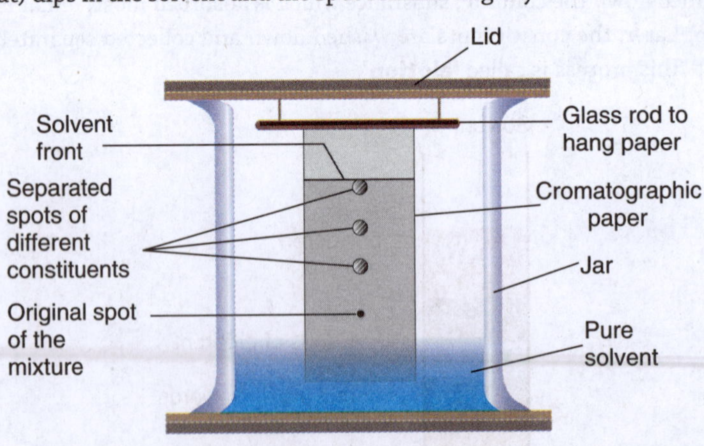

Fig. 6.2. Paper chromatography.

The solvent rises up the paper due to capillary action and the components of the mixture rise up at different rates and thus get separated from one another as shown in Fig. 6.2.

This type of paper chromatography in which the solvent rises up is called **'Ascending paper chromatography.'** Alternatively, the solvent may be taken on the top in a container and be allowed to come down in which case it is termed as **'Descending paper chromatography'.**

R_f Values

It represents rentention factor or ratio of fronts. It may be defined as *the ratio of the distance travelled by the component from the origin or point of application to the distance moved up by the solvent from the same point.*

$$R_f = \frac{\text{Distance travelled by the solute from the original line}}{\text{Distance travelled by the solvent from the original line}}$$

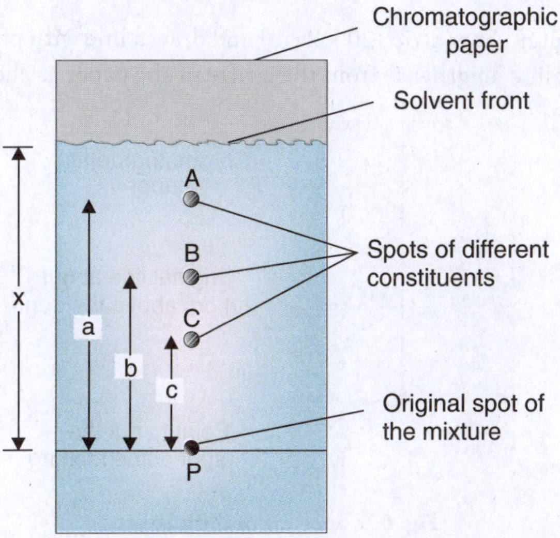

Fig. 6.3. Calculation of R_f values.

For example, the R_f values of substances A, B and C will be as given under :

$$R_f \text{ value for A} = \frac{a}{X}$$

$$R_f \text{ value for B} = \frac{b}{X}$$

$$R_f \text{ value for C} = \frac{c}{X}$$

Different substance possess different R_f values. R_f depends upon a number of factors :

(*i*) *Nature of the substance.*

(*ii*) *Nature of the solvent.*

(*iii*) *Temperature.*

(*iv*) *Presence of impurities.*

(*v*) *Quality of the filter paper used.*

EXPERIMENT 6.1

AIM: ***To separate the coloured components present in the mixture of red and blue inks by ascending paper chromatography and find their R_f values.***

APPARATUS

Gas jar, glass rod, filter paper strip (Whatman No. 1 filter paper), jar cover and a fine capillary tube.

REQUIREMENT

A mixture of red and blue inks, alcohol and distilled water.

PROCEDURE

1. Take a Whatman filter paper strip (20 × 2 cm) and draw a line with pencil above 3 cm from one end. Draw another line lengthwise from the centre of the paper as shown in Fig. 6.4.

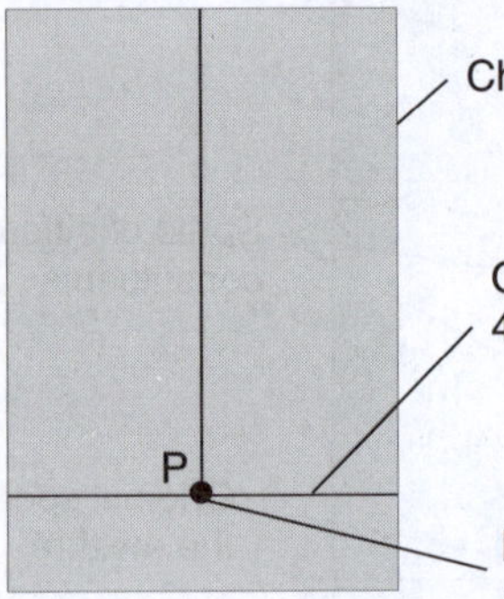

Fig. 6.4. Spotting of the mixture.

2. With the help of fine capillary tube, put a drop of the mixture of red and blue inks at the point P. Let it dry in air. Put another drop on the same spot and dry again. Repeat 2–3 times, so that the spot is rich in the mixture.
3. Suspend the filter paper vertically in a gas jar containing the solvent (eluent) with the help of a glass rod in such a way that the pencil line (and the spot) remains about 2 cm above the solvent level (50% alcohol + distilled water).
4. Cover the jar and keep it undisturbed. Notice the rising solvent along with the mixture of red and blue inks. After the solvent has risen about 15 cm you will notice two different spots of blue and red colours on the filter paper.
5. Take the filter paper out of the jar and mark the distance that the solvent has risen on the paper with a pencil. This is called the solvent front.
6. Dry the paper. Put pencil marks in the centre of the blue and red spots.
7. Measure the distance of the two spots from the original line and the distance of the solvent from the original line.

8. Calculate the R_f values of the blue and red inks by using the formula :

$$R_f = \frac{\text{Distance travelled by the blue or red ink from the point of application}}{\text{Distance travelled by the solvent from the original line}}$$

OBSERVATIONS AND CALCULATIONS

Substance	*Distance travelled by different components*	*Distance travelled by solvent*	*R_f Value*
Red ink + Blue ink	A cm (Red Ink) B cm (Blue Ink)	X cm X cm	A/X B/X

PRECAUTIONS

1. Use good quality pencil for drawing the reference line so that the mark does not dissolve in the solvent in which the chromatography is carried out.
2. Always make use of a fine capillary tube.
3. Keep the jar undisturbed and covered during the experiment.
4. A spot should be small and rich in mixture.
5. Allow the spot to dry before putting the strip in the jar.
6. Keep the strip erect. Do not let it to be curled.
7. Do not allow the spot to dip in the solvent.

EXPERIMENT 6.2

AIM: ***To separate the coloured components present in the given grass/flower by ascending paper chromatography and determine their R_f values.***

In this experiment, crush fresh flowers or grass in a mortar and extract the juice with acetone. Use this solution for making the spot.

Proceed as in Expt. 6.1.

OBSERVATIONS AND CALCULATIONS

Colour of the spot	*Distance travelled by the spot from the original line*	*Distance travelled by the solvent from the original line*	*R_f Value*
Green (*Chlorophyll*)	A cm	X cm	A/X
Yellow *Xanthophyll*)	B cm	X cm	B/X
Red (*Carotene*)	C cm	X cm	C/X

EXPERIMENT 6.3

AIM: ***To separate Co^{+2} and Ni^{2+} ions present in the given mixture by using ascending paper chromatography and determine their R_f values.***

APPARATUS

Gas jar, glass rod, filter paper strip (Whatman No. 1 filter paper), jar cover and a fine capillary tube.

CHEMICALS REQUIRED

Sample solution containing cobalt (II) and nickel (II) ions, acetone, concentrated aqueous ammonia, Rubeanic acid spray reagent.

PROCEDURE

Same as in experiment 6.1. Use a mixture of acetone (90%), concentrated hydrochloric acid (5%) and water (5%) as eluent.

After elution and drying, place the paper in a large, dry, covered beaker containing a smaller beaker of concentrated aqueous ammonia. After about two minutes, remove the paper and spray it, on both sides, with rubeanic acid reagent. Allow it to dry. Nickel becomes visible as blue purple band while cobalt becomes visible as yellow orange band.

Evaluate R_f values of the two ions.

OBSERVATIONS AND CALCULATIONS

Colour of the spot	*Distance travelled by different components*	*Distance travelled by solvent*	*R_f value*
Blue purple (Ni^{2+})	A cm	X cm	A/X
Yellow orange (Co^{2+})	B cm	X cm	B/X

RESULT

R_f value of Ni^{2+} = ...

R_f value of Co^{2+} = ...

The above experiment can be carried by using a mixture of

(*i*) Iron (II) and cobalt (II)

(*ii*) Iron (II) and nickel (II)

(*iii*) Copper (II) and iron (II)

(*iv*) Copper (II) and nickel (II)

(*v*) Iron (II) and zinc (II)

(*vi*) Lead (II) and Cadmium (II).

VIVA QUESTIONS WITH ANSWERS

Q. 1. What is chromatography ?

Ans. It is a technique for separation of components of a mixture and purification of compounds. It is based on differential migration of the various components of a mixture through a stationary phase under the influence of a moving phase.

Q. 2. What is the principle of chromatographic process ?

Ans. It is based on the differential migration of the individual components of a mixture through a stationary phase under the influence of a moving phase.

Q. 3. What type of solvents are generally employed in chromatography ?

Ans. Generally solvents having low viscosities are employed in chromatography. This is due to the fact that the rate of flow of a solvent varies inversely as its viscosity.

Q. 4. Name some chromatographic techniques.

Ans. Paper chromatography, column chromatography, thin layer chromatography, gas chromatography.

Q. 5. What are the moving and stationary phases in paper chromatography ?

Ans. Water absorbed on cellulose constituting the paper serves as the stationary phase and organic solvent as moving phase.

Q. 6. What is meant by the term developing in chromatography ?

Ans. During chromatography, if the components to be separated are colourless, then these separated components on chromatogram are not visible. Their presence is detected by *development*, which involves spraying a suitable reagent (called *developing reagent*) on the chromatogram, or placing the chromatogram in iodine chamber, when various components become visible. *This process is called* **developing of chromatogram.**

Q. 7. How does the liquid rise through the filter paper ?

Ans. Through capillary action.

Q. 8. What is meant by the term R_f value ?

Ans. R_f (retention factor) value of a substance is defined as the ratio of the distance moved up by the solute from the point of its application to the distance moved up by the solvent from the same point.

Q. 9. On what factors does the R_f value of a compound depend ?

Ans. (*i*) Nature of the compound.

(*ii*) Nature of the solvent.

(*iii*) Temperature.

Q. 10. Name the scientist who introduced chromatographic technique.

Ans. Russian botanist **M. Tswett** (1906).

Q. 11. What are the advantages of chromatography over other techniques ?

Ans. (*i*) It can be used for a mixture containing any number of components.

(*ii*) Very small quantities of the substances can be effectively detected and separated from a mixture.

Q. 12. What is loading (or spotting) ?

Ans. The application of the mixture as a spot on the original line on the filter paper strip or addition of mixture to the column, is called *loading* (or *spotting*).

Q. 13. What are the essential characteristics of the substance used as a developer ?

Ans. (*i*) It should be volatile.

(*ii*) It should impart colour to the different spots.

(*iii*) It should not react with various compounds which are being separated.

CHAPTER 7

Preparation of Inorganic Compounds

A double salt is a substance obtained by the combination of two different salts which crystallise together as a single substance but ionize as two distinct salts when dissolved in water. The constituent salts are always taken in some definite molecular proportions. *Alums* are common examples of double salts.

Alums *are double sulphates having general formula,*

$$X_2SO_4.M_2(SO_4)_3.24H_2O$$

where, X = monovalent cation such as Na^+, K^+, NH_4^+ etc.

M = trivalent cation such as Al^{3+}, Cr^{3+}, Fe^{3+} etc.

Some important alums and their names are given below :

Potash Alum	:	$K_2SO_4.Al_2(SO_4)_3.24H_2O$
Chrome Alum	:	$K_2SO_4.Cr_2(SO_4)_3.24H_2O$
Soda Alum	:	$Na_2SO_4.Al_2(SO_4)_3.24H_2O$
Ferric Alum	:	$(NH_4)_2SO_4.Fe_2(SO_4)_3.24H_2O$

Alums are *isomorphous* crystalline solids which are soluble in water. Due to hydrolysis, their aqueous solutions have acidic character.

Another example of double salts is **Mohr's salt.** Its formula is $FeSO_4.(NH_4)_2SO_4.6H_2O$. It is used as primary standard in volumetric analysis. Its crystals do not lose water of crystalisation by efflorescence nor it is oxidised in air. It is stable salt unlike green vitriol ($FeSO_4.7H_2O$) which gets oxidised by air.

Before we discuss preparation of some of these double salts, let us first review the process of crystallisation.

7.01. PROCESS OF CRYSTALLISATION

The process of crystallisation involves following steps :

1. Preparation of Solution of the Impure Sample

1. Take a clean beaker (250 ml) and add powdered impure sample under consideration in it (~ 6.0 gm).
2. Add distilled water (25–30 ml) and stir contents gently with the help of a glass rod giving circular motion as shown in Fig. 7.1.
3. The solution in the beaker is heated (60°–70°C) on a wire gauze (Fig. 7.2).

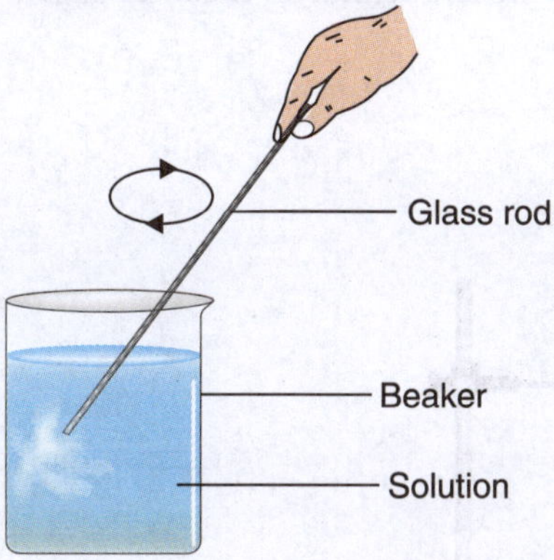

Fig. 7.1. Stirring with a glass rod.

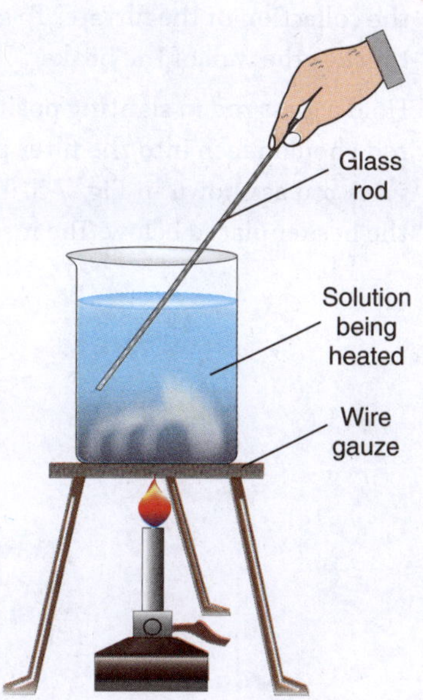

Fig. 7.2. Heating of solution.

4. Stir the solution continuously and add more of impure substance till no more of it dissolves.

2. Filtration of Hot Solution

1. Take a circular filter paper. First fold it one-half, then fold it one-fourth as shown in Fig. 7.3. Open the filter paper, three folds on one side and one fold on the other side to get a cone (Fig. 7.3).

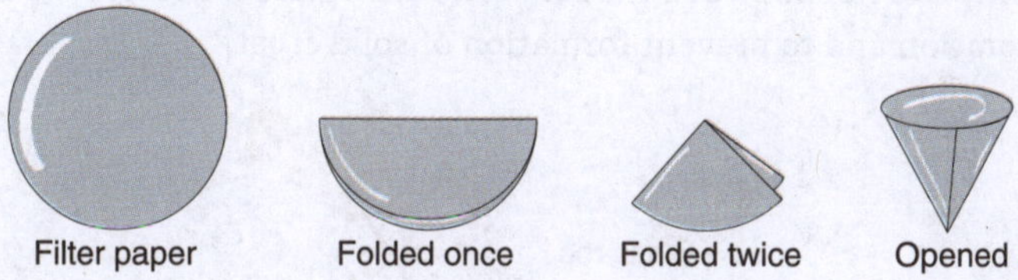

Fig. 7.3. Making a cone.

2. Take a funnel and fit the filter paper cone into the funnel so that the upper half of the cone fits well into the funnel but lower part remains slightly away from the funnel.
3. Wet the filter paper cone with a spray of water from a wash bottle pressing the upper part of the filter paper cone gently against the wall of the funnel with the thumb (Fig. 7.4).
4. Place the funnel on a funnel stand and place a clean beaker below the funnel for

Fig. 7.4. Fitting and wetting of filter paper cone.

the collection of the filtrate. To avoid splashing of the filtrate, adjust the funnel so that its stem touches the wall of the beaker.

5. Hold a glass rod in slanting position in your hand or with a precaution that the lower end of the rod should reach into the filter paper cone but it does not touch it. Pour the solution along the glass rod as shown in Fig. 7.5. The filtrate passes through the filter paper and is collected into the beaker placed below. The insoluble impurities are left behind on the filter paper.

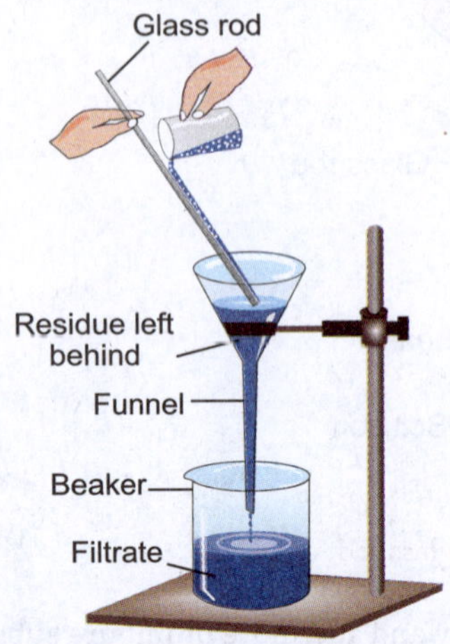

Fig. 7.5. Removing insoluble impurities by filtration.

3. Concentration of Filtrate

1. Transfer the filtrate to a china dish and place the dish on wire gauze, kept over a tripod stand and heat it gently (Do not boil). Stir the solution with a glass rod (Fig. 7.6). This is done to ensure uniform evaporation and to prevent formation of solid crust.

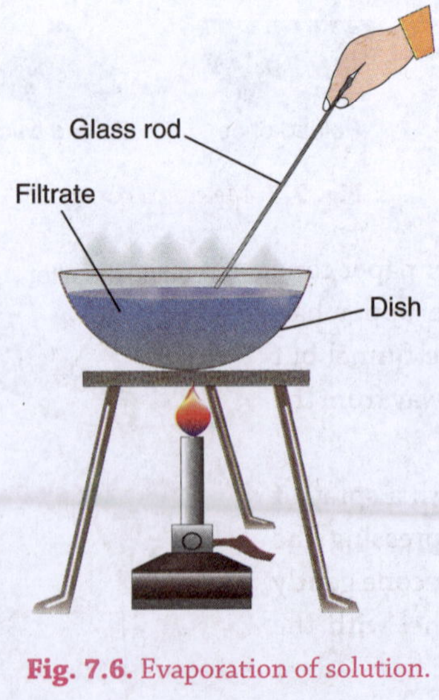

Fig. 7.6. Evaporation of solution.

2. When the volume of the solution is reduced to one-half, take out a drop of the concentrated solution on one end of glass rod and cool it by blowing air (Fig. 7.7). Formation of thin crust indicates that the **crystallisation point** has reached.

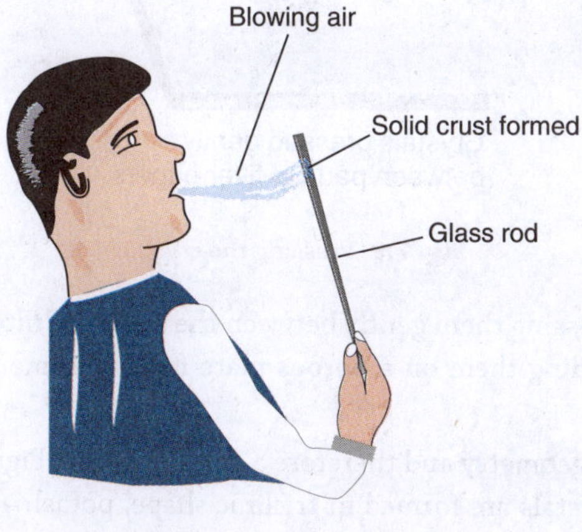

Fig. 7.7. Checking the crystallisation point.

3. Stop heating by removing the burner.

4. Cooling the Concentrated Solution

1. Pour the concentrated solution into a *crystallising dish*. (It is a thin walled shallow glass dish with a flat bottom and vertical sides. It has a spout to pour off the mother liquor).
2. Cover the dish with a watch glass and keep it undisturbed (Fig. 7.8).
3. As the solution cools, crystals separate out. The concentrated solution is cooled slowly for better yield of the crystals.

Sometimes the china dish containing the concentrated solution is cooled by placing on a beaker filled to the brim with cold water. Cooling may also be done by keeping the china dish in open air depending upon the weather conditions.

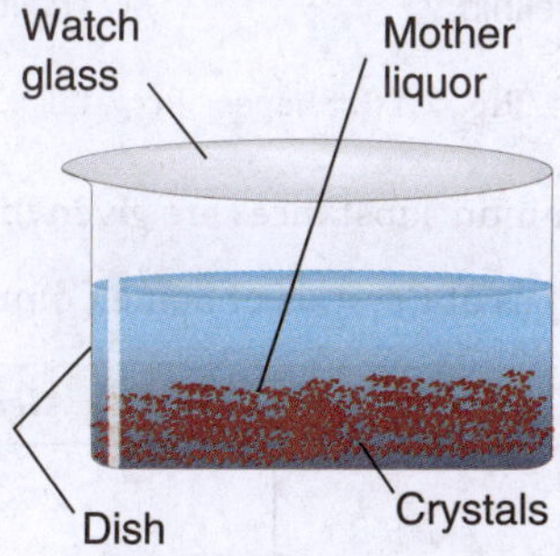

Fig. 7.8. Cooling in crystallising dish.

5. Separation and Drying of Crystals

1. Decant off the mother liquor and wash the crystals with a thin stream of cold water with the help of a wash bottle.

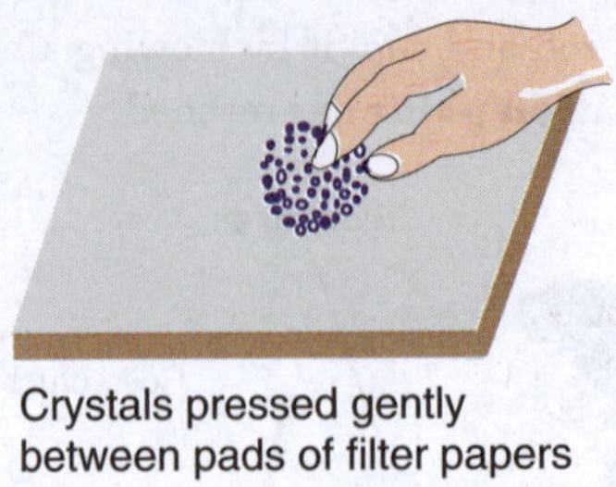

Fig. 7.9. Pressing the crystals.

2. Dry the crystals by pressing them gently between the sheets of filter paper Fig. 7.9. The crystals can be dried by spreading them on a porous plate for sometime or by placing the crystals in vacuum desiccator.

Crystals have definite geometry and therefore a definite shape. Figure 7.10 shows some of these shapes. Copper sulphate crystals are formed in triclinic shape, potash alum comes out in octahedral geometry. Potassium nitrate crystals are needle like and ferrous sulphate have monoclinic shape.

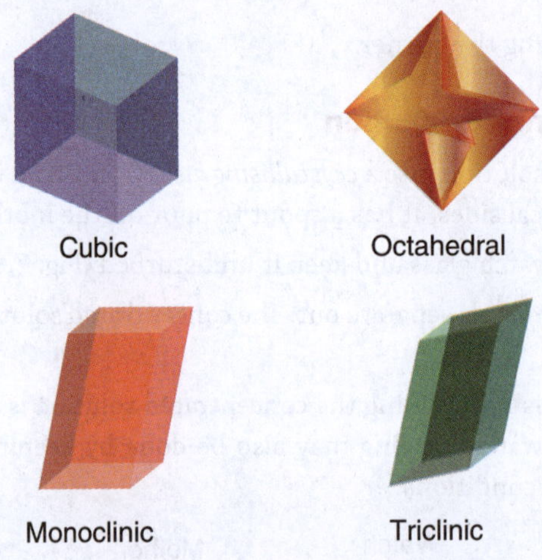

Fig. 7.10. Shapes of crystals.

Shapes of crystals of some common substances are given in Table 7.1.

Table 7.1. Shapes of Crystals of Some Common Substances

Substance	***Geometry or shape of crystal***
1. Blue vitriol, $CuSO_4.5H_2O$	Triclinic
2. Green vitriol, $FeSO_4.7H_2O$	Monoclinic
3. Potassium nitrate, KNO_3	Rhombic (Needle-like)
4. Potash alum, K_2SO_4, $Al_2(SO_4)_3$,$24H_2O$	Octahedral
5. Sodium chloride	Cubic
6. Washing soda, $Na_2CO_3.10H_2O$	Monoclinic

EXPERIMENT 7.1

AIM: ***To prepare a pure sample of ferrous ammonium sulphate (Mohr's salt) [$FeSO_4 . (NH_4)_2SO_4 . 6H_2O$].***

THEORY

Mohr's salt is prepared by dissolving an equimolar mixture of hydrated ferrous sulphate and ammonium sulphate in water containing a little of sulphuric acid, and then subjecting the resulting solution to crystallisation when **light green** crystals of ferrous ammonium sulphate. $FeSO_4 . (NH_4)_2SO_4 . 6H_2O$ separate out.

$$\underset{\substack{\text{Ferrous sulphate} \\ 278}}{FeSO_4.7H_2O} + \underset{\substack{\text{Ammonium sulphate} \\ 132}}{(NH_4)_2SO_4} \longrightarrow \underset{\substack{\text{Mohr's salt} \\ 392}}{FeSO_4.(NH_4)_2SO_4.6H_2O} + H_2O$$

REQUIREMENTS

Two beakers (250 ml), china-dish, funnel, funnel stand, glass rod, wash-bottle, tripod stand and wire-gauze.

Ferrous sulphate crystals, ammonium sulphate crystals, dilute sulphuric acid and ethyl alcohol.

PROCEDURE

1. Take a 250 ml beaker and wash it with water. Transfer 7.0 g ferrous sulphate and 3.5 g ammonium sulphate crystals to it. Add about 2–3 ml of dilute sulphuric acid to prevent the hydrolysis of ferrous sulphate.
2. In another beaker boil about 20 ml of water for about 5 minutes to expel dissolved air.
3. Add the boiling hot water to the contents in the first beaker in small instalments at a time. Stir with a glass rod until the salts have completely dissolved.
4. Filter the solution to remove undissolved impurities and transfer the filtrate to a china-dish.
5. Heat the solution in the china-dish for some time to concentrate it to the crystallisation point.
6. Place the china-dish containing saturated solution over a beaker full of cold water. On cooling crystals of Mohr's salt separate out.
7. Decant off the mother liquor quickly. Wash the crystals in the china-dish with a small quantity of alcohol to remove any sulphuric acid sticking to the crystals.
8. Dry the crystals by placing them between filter paper pads.

OBSERVATIONS

Weight of crystals obtained = ... g

Expected yield = ... g

Colour of the crystals = ...

Shape of the crystals = ...

Note: *The crystals of Mohr's salt are **monoclinic** in shape.*

PRECAUTIONS

1. Cool the solution slowly to get good crystals.
2. Do not disturb the solution while it is being cooled.
3. Do not heat the solution for a long time as it may oxidize ferrous ions to ferric ions.

EXPERIMENT 7.2

AIM: ***To prepare a pure sample of potash alum (Fitkari), $[K_2SO_4.Al_2(SO_4)_3.24H_2O]$.***

THEORY

Potash alum is prepared by dissolving an equimolar mixture of hydrated aluminium sulphate and potassium sulphate in minimum amount of water containing a little of sulphuric acid and then subjecting the resulting solution to crystallisation, when octahedral crystals of potash alum separate out.

$$\underset{\substack{\text{Potassium}\\\text{sulphate}\\174}}{K_2SO_4} + \underset{\substack{\text{Aluminium}\\\text{sulphate}\\666}}{Al_2(SO_4)_3.18H_2O} + 6H_2O \longrightarrow \underset{\text{Potash alum}}{K_2SO_4.Al_2(SO_4)_3.24H_2O.}$$

REQUIREMENTS

Two beakers (250 ml), china-dish, funnel, funnel stand, glass rod, wash-bottle, tripod stand and wire-gauze.

Potassium sulphate, aluminium sulphate and dil. sulphuric acid.

PROCEDURE

1. Take a 250 ml beaker. Wash it with water and then transfer 2.5 g potassium sulphate crystals to it. Add about 20 ml of water. Stir to dissolve the crystals. Warm if required.
2. Take the other 250 ml beaker, wash it with water and then transfer 10 g aluminium sulphate crystals to it. Add about 20 ml of water and 1 ml of dilute sulphuric acid to prevent hydrolysis of aluminium sulphate. Heat for about 5 minutes. If milkiness still persists, filter the solution.
3. Mix the two solutions in a china-dish and place the china-dish on a wire-gauze placed over a burner. Stir the solution with a glass rod. *Concentrate the solution till the crystallisation point is reached.* Place the dish over a beaker containing cold water.
4. Soon the crystals of potash alum separate out. Decant off the mother liquor and wash the crystals with a small quantity of ice-cold water.
5. Dry the crystals by placing them between filter paper pads or by spreading them over porous plate.

OBSERVATIONS

Weight of crystals obtained	= ...... g
Expected yield	= ...... g
Colour of the crystals	= ......
Shape of the crystals	= ......

Note: *The crystals of potash alum are* ***octahedral*** *in shape.*

PRECAUTIONS

1. Cool the solution slowly to get good crystals.
2. Do not disturb the solution while it is being cooled.

EXPERIMENT 7.3

AIM: ***To prepare a pure sample of the complex potassium trioxalatoferrate(III), $K_3[Fe(C_2O_4)_3] . 3H_2O$.***

THEORY

The complex potassium trioxalatoferrate(III) can be prepared by dissolving freshly prepared ferric hydroxide in a solution of potassium oxalate and oxalic acid.

$$\underset{162.7}{FeCl_3} + 3KOH \longrightarrow \underset{107}{Fe(OH)_3} + 3KCl$$

$$\underset{2 \times 107}{2Fe(OH)_3} + \underset{3 \times 126}{3\begin{array}{c}COOH\\|\\COOH\end{array} . 2H_2O} \longrightarrow \underset{376}{Fe_2(C_2O_4)_3} + 12H_2O$$

$$\underset{376}{Fe_2(C_2O_4)_3} + \underset{3 \times 184}{3\begin{array}{c}COOK\\|\\COOK\end{array} H_2O} \longrightarrow \underset{2 \times 491}{2K_3[Fe(C_2O_4)_3] . 3H_2O}$$

REQUIREMENTS

Three beakers (250 ml), china-dish, funnel, funnel stand, glass rod, wash-bottle, tripod stand and wire-gauze.

Ferric chloride, oxalic acid hydrated, potassium oxalate and potassium hydroxide.

PROCEDURE

1. Dissolve 3.5 g of anhydrous ferric chloride 50 ml of distilled water in a 250 ml beaker.
2. In another beaker dissolve 4 g of potassium hydroxide in 50 ml of water.
3. Add KOH solution to $FeCl_3$ solution in small portions with constant stirring. Filter the precipitates of ferric hydroxide so formed through a buchner funnel. Wash the ppt. with distilled water.
4. In another beaker (250 ml) take 4 g of hydrated oxalic acid and 5.5 g of hydrated potassium oxalate. Add about 100 ml of water and stir thoroughly to get a clear solution.
5. Add the freshly prepared $Fe(OH)_3$ ppt. in small amounts to the above solution with constant stirring. The ppt. get dissolved. If ppt. does not dissolve then warm it and leave the contents for sometime.
6. Filter and transfer the filtrate to china dish and heat on a wire-gauze to the crystallisation point.
7. Now place the china dish on a beaker full of cold water and keep it aside for crystallisation. China dish should be covered with a black paper as the complex is sensitive to light.

8. Decant off the mother liquor, wash the crystals with a small amount of ethyl alcohol and dry them between the folds of filter paper.
9. Find out the weight of the crystals.

OBSERVATIONS

Weight of the crystals obtained = ... g

Colour of the crystals is ....

PRECAUTIONS

1. Do not concentrate the solution too much.
2. Let the concentrated solution cool slowly and undisturbed to get large crystals.

VIVA QUESTIONS WITH ANSWERS

Q. 1. Define the term 'crystallisation'.

Ans. The substances when present in well-defined geometrical shapes are called crystals. These are formed when a hot saturated solution of the salt is allowed to cool slowly and undisturbed. *The process of obtaining crystals is termed as crystallisation.*

Q. 2. What is meant by equimolar proportions ?

Ans. Proportion of the substances in the ratio of their molecular masses, *i.e.*, 1 : 1 mole ratio.

Q. 3. Why is the hot saturated solution not cooled suddenly ?

Ans. If the solution is cooled suddenly, crystals of smaller size are formed. By allowing saturated solution to cool slowly, crystals grow in size.

Q. 4. What is the term 'seeding' ?

Ans. Sometimes on cooling the saturated solution, crystallisation does not occur. A crystal of same substance is placed in the saturated solution which induces crystallisation. This process is known as seeding. It helps in quick separation of crystals from saturated solution.

Q. 5. Does lithium sulphate combine with aluminium sulphate to form alum ?

Ans. No, lithium ion being very small in size does not form alums.

Q. 6. What is green vitriol ?

Ans. It is hydrated ferrous sulphate ($FeSO_4.7H_2O$).

Q. 7. What is mother liquor ?

Ans. The liquid left behind after the separation of crystals from a saturated solution is known as mother liquor. It contains soluble impurities.

Q. 8. What are alums ?

Ans. Alums are double sulphates having general formula $X_2SO_4.M_2(SO_4)_3.24H_2O$, where X = monovalent cation such as Na^+, K^+ etc. and M = trivalent cation such as Al^{+3}, Cr^{+3}, etc.

Q. 9. In the preparation of Mohr's salt can concentrated H_2SO_4 be used in place of dilute H_2SO_4 ?

Ans. No, because it would oxidize ferrous ions to ferric ions.

Q. 10. What is the action of heat on potash alum ?

Ans. It loses water of crystalization and becomes light and fluffy.

Q. 11. Give the names of some alums where cations are other than Al^{3+}.

Ans. Ferric alum, $(NH_4)_2SO_4 \cdot Fe_2(SO_4)_3 \cdot 24H_2O$; Chrome alum, $K_2SO_4 \cdot Cr_2(SO_4)_3 \cdot 24H_2O$.

Q. 12. What are isomorphous substances ?

Ans. The substances having similar crystal structure are known as isomorphous.

Q. 13. Why is dilute sulphuric acid added to the solution during the preparation of Mohr's salt crystals ?

Ans. It prevents hydrolysis of ferrous sulphate.

Q. 14. What are the uses of potash alum ?

Ans. It is used for purification of impure water. It is also used to stop bleeding from a wound and as mordant in dyeing industry.

Q. 15. Why is water, used for the preparation of Mohr's salt solution, boiled for 5 minutes ?

Ans. It is done in order to expel dissolved oxygen from the water which otherwise would oxidize ferrous salt to ferric salt.

Q. 16. Why is dilute sulphuric acid added during the preparation of aluminium sulphate solution ?

Ans. To prevent the hydrolysis of aluminium sulphate.

Q. 17. How does potash alum help in purification of water ?

Ans. When potash alum is added to impure water, it causes the coagulation of colloidal impurities present in water. The precipitated impurities can be removed by filtration or decantation.

Q. 18. How does potash alum help in stopping bleeding ?

Ans. Blood is a negatively charged sol, in the presence of potash alum it gets coagulated.

Q. 19. Is aqueous solution of potash alum acidic or basic ?

Ans. It is acidic, it turns blue litmus paper red. The solution is acidic due to hydrolysis of aluminium sulphate.

Q. 20. What is the geometry of the complex $K_3[Fe(C_2O_4)_3] \cdot 3H_2O$?

Ans. Octahedral.

Q. 21. Why is the complex $K_3[Fe(C_2O_4)_3] \cdot 3H_2O$ paramagnetic ?

Ans. It is paramagnetic due to the presence of five unpaired electrons in 3*d*-orbitals of Fe (III).

Q. 22. What is the IUPAC name of the complex $K_3[Fe(C_2O_4)_3] \cdot 3H_2O$?

Ans. Potassium trioxalatoferrate (III)-3-water.

Q. 23. What is the difference between a complex and a double salt ?

Ans. In complex salt, the properties of all individual ions of the constituent salts may not be exhibited. In case of double salts properties of the ions of the constituent salts are exhibited in solution.

Q. 24. What is the difference between iron compounds given below ?

$K_4[Fe(CN)_6]$ and $FeSO_4.(NH_4)_2SO_4.6H_2O$.

Ans. $K_4[Fe(CN)_6]$ is a coordination complex whereas $FeSO_4.(NH_4)_2SO_4.6H_2O$ is a double salt.

Q. 25. What is the coordination number of iron is potassium trioxalatoferrate (III) ?

Ans. Six

Q. 26. List two examples of bidentate ligands other than oxalate ion.

Ans. *(i)* Ethylene diamine (en), $H_2N{-}CH_2{-}CH_2{-}NH_2$

(ii) Glycinate, $H_2N{-}CH_2{-}COO^-$.

Q. 27. Why does the compound, potassium trioxaloferrate (III), not give tests for ferric ions ?

Ans. Because it contains Fe(III) as complex ion, $[Fe(C_2O_4)_3]^{3-}$ and not as Fe^{3+} ions.

Q. 28. What are chelates ?

Ans. Multidentate ligands are known as chelates.

CHAPTER 8

Preparation of Organic Compounds

In this chapter, we shall discuss the preparations of some organic compounds.

EXPERIMENT 8.1

AIM: ***To prepare acetanilide from aniline.***

THEORY

Acetanilide is prepared by acetylating aniline with acetic anhydride in the presence of glacial acetic acid. The chemical equation can be written as :

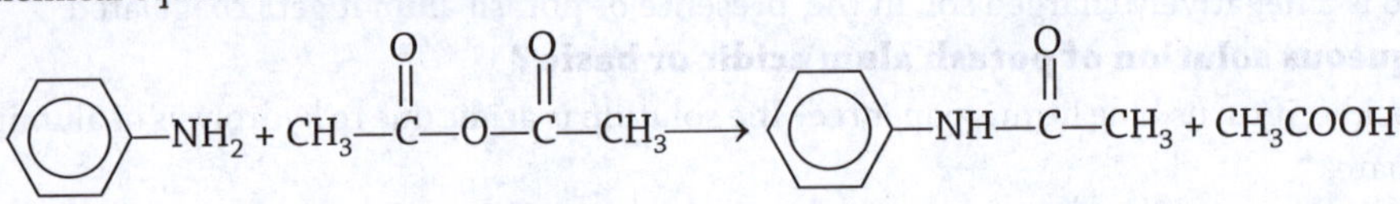

Aniline Acetic anhydride Acetanilide

APPARATUS

Round bottom flask (100 ml), water condenser, wire-gauze, glass rod, tripod stand, burner, iron-stand, clamp, measuring cylinder, etc.

CHEMICALS REQUIRED

Aniline	= 5 ml
Acetic anhydride	= 5 ml
Glacial acetic acid	= 5 ml.

PROCEDURE

1. Take 5 ml of acetic anhydride in a clean dry 100 ml conical flask and add 5 ml of glacial acetic acid and shake the contents thoroughly.
2. To this mixture taken in the flask, add 5 ml of aniline and fit a water condenser.
3. Place the flask on a wire-gauze placed on a tripod stand as shown in Fig. 8.1.
4. Boil the mixture for 10–15 minutes.
5. Allow the mixture to cool. Detach the condenser and pour the liquid into about 150 ml ice-cold water contained in a beaker. During addition, stir vigorously the contents of the beaker with the help of glass rod.

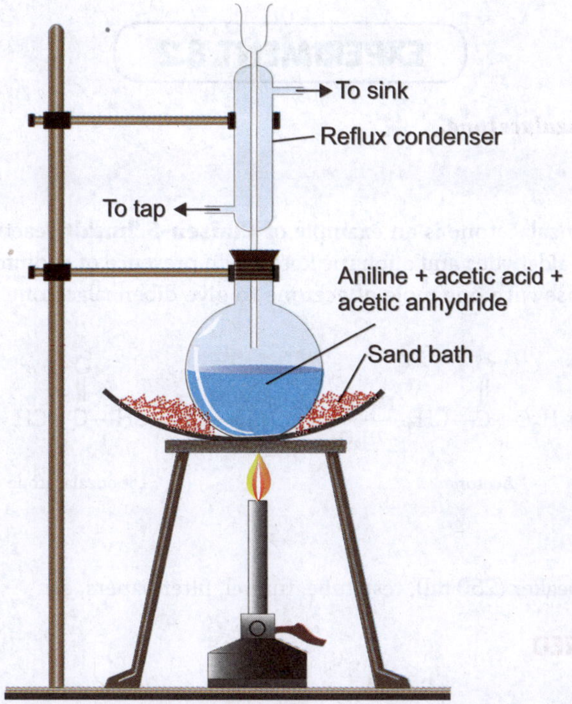

Fig. 8.1. Preparation of acetanilide.

6. Filter the white precipitates which separate out and wash with cold water.
7. Recrystallise from hot water containing a few drops of ethyl alcohol. Weigh the crystals and record the yield.
8. Determine the melting point of the compound.

RESULT

Weight of acetanilide obtained = ... g

Melting point of acetanilide = ... °C

Note: *Acetanilide has white flaky crystals. Its melting point in 113°C.*

PRECAUTIONS

1. Freshly distilled aniline should be used in order to get good results or **small amount of zinc can be added in the reaction mixture.** Zinc reduces the coloured impurities in the aniline and also prevents its oxidation during the reaction.
2. Prolonged heating and use of excess of acetic anhydride should be avoided.
3. Reaction mixture should first be cooled and then poured in ice-cold water, otherwise hydrolysis of acetanilide may take place.

EXPERIMENT 8.2

AIM: ***To prepare dibenzalacetone.***

THEORY

The preparation of dibenzalacetone is an example of **Claisen-Schmidt** reaction. This reaction takes place between aromatic aldehydes and aliphatic ketones in presence of sodium hydroxide. Two moles of benzaldehyde condense with one mole of acetone to give dibenzalacetone. The chemical equation can be written as :

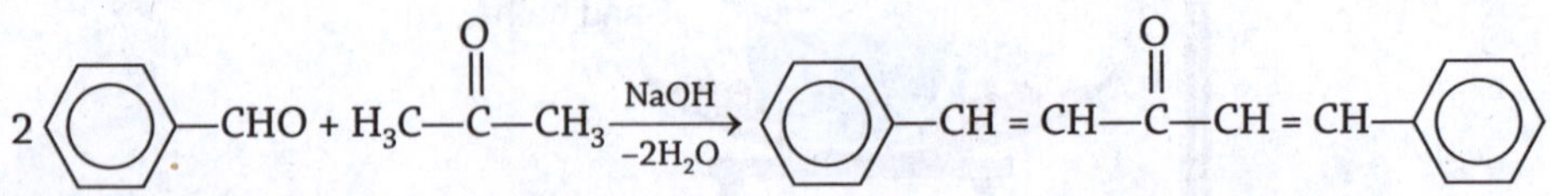

Benzaldehyde　　Acetone　　Dibenzalacetone

APPARATUS

Conical flask (100 ml), beaker (250 ml), test-tube, funnel, filter papers, etc.

CHEMICALS REQUIRED

Benzaldehyde	= 2.5 ml
Acetone	= 1.0 ml
10% NaOH solution	= 5 ml
Rectified spirit	= 25 ml

PROCEDURE

1. Take a conical flask (100 ml) and add 2.5 ml benzaldehyde, 1.0 ml of acetone and 25 ml of methylated spirit. Cork the flask and shake to obtain a clear solution.
2. Take 5 ml of 10% NaOH solution in a test tube and add this to conical flask drop by drop with shaking of the flask. Maintain the temperature of the reaction mixture between 20–25°C during addition of sodium hydroxide solution.
3. Cork the flask again and shake vigorously for about 10 minutes, releasing pressure from time to time.
4. Allow it to stand for about 20 minutes at room temperature and then cool in ice water for a few minutes.
5. Filter the yellow coloured solid and wash it with water to remove traces of alkali.
6. *Recrystallization of dibenzalacetone.*
   Dissolve the above yellow coloured crude solid in minimum amount of hot rectified spirit and then allow it to cool slowly. Pale yellow crystals of dibenzalacetone separate out. Filter the crystals and dry.
7. Weigh and record its yield and melting point.

RESULT

Weight of dibenzalacetone obtained = ... g
Melting point of dibenzalacetone is ...... °C
Note: *(Approximate expected yield of dibenzalacetone is 1.5 g)*
The melting point of dibenzalacetone is 112°C.

PRECAUTIONS

1. Always use freshly distilled benzaldehyde or the sample from a freshly opened bottle.
2. Add NaOH dropwise to the reaction mixture with constant shaking and maintaining the temperature around 20°C.
3. Use minimum amount of rectified spirit to dissolve crude sample for crystallisation.

EXPERIMENT 8.3

AIM: ***To prepare p-nitroacetanilide from acetanilide.***

THEORY

The nitration of aniline is difficult to carry out with nitrating mixture (a mixture of conc. H_2SO_4 and conc. HNO_3) since —NH_2 group gets oxidised, which is not required. So the amino group is first protected by acylation to form acetanilide which is then nitrated to give *p*-nitroacetanilide as the major product and *o*-nitroacetanilide as minor product. Recrystallisation from ethanol readily removes the more soluble *ortho*-compound and the pure *p*-nitroacetanilide is obtained. The chemical equation can be written as :

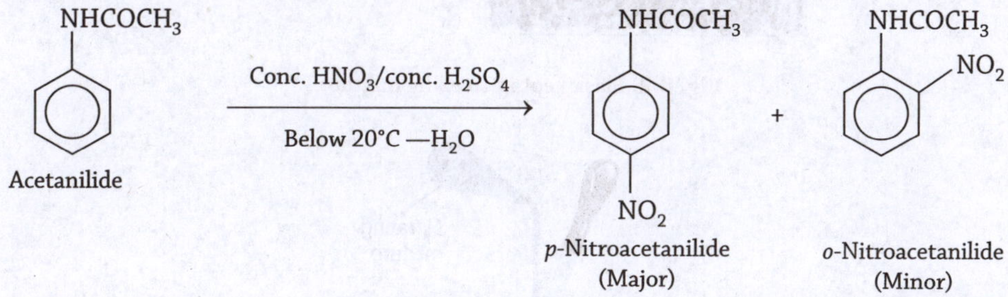

APPARATUS

Conical flask (100 ml), beaker (250 ml), measuring cylinder (100 ml), funnel, glass rod, test tube, filter papers, etc.

CHEMICALS

Acetanilide	= 5 g
Glacial acetic acid	= 5 ml
Conc. H_2SO_4	=10 ml
Fuming HNO_3	= 2 ml
Methylated spirit	= 20 ml.

PROCEDURE

1. Take a 100 ml conical flask and add 5 g of powdered acetanilide in it. Add 5 ml of glacial acetic acid and stir the mixture by the use of glass rod.
2. Place 2 ml of fuming nitric acid in a clean test-tube and cool it in a freezing mixture (ice + salt) taken in a beaker. Carefully add drop by drop 2 ml of conc. sulphuric acid with constant shaking and cooling.

3. Add the remaining 8 ml of conc. H_2SO_4 drop by drop (with cooling under tap water) to the conical flask containing acetanilide and glacial acetic acid. Place the conical flask in a freezing mixture (Fig. 8.2). Stir the contents and wait until the temperature becomes less than 5°C.
4. To the cooled contents in the flask add nitrating mixture prepared in step (2) drop by drop with constant stirring (Fig. 8.3). During addition temperature of the mixture should not rise above 10°C. This operation should take about 15 minutes.
5. Remove the conical flask from the freezing mixture and allow it to stand for 30 minutes at room temperature.
6. Pour the contents of the flask on the crushed ice taken in a beaker. Stir it and filter the crude product. Wash thoroughly with cold water to remove acid.

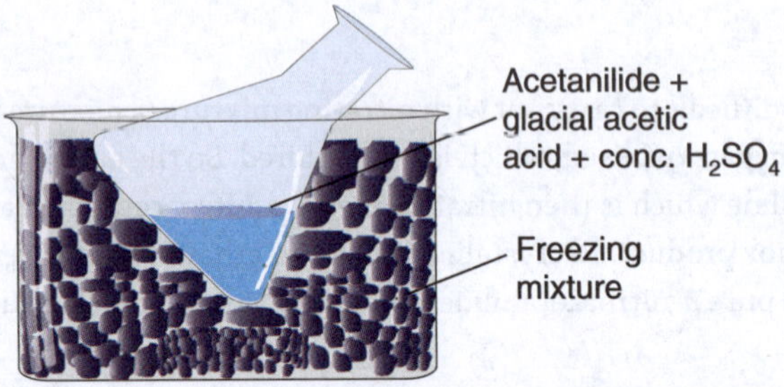

Fig. 8.2. Flask kept in freezing mixture.

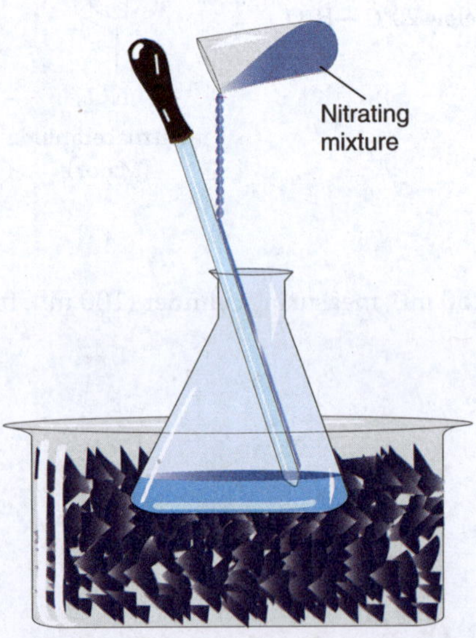

Fig. 8.3. Preparation of *p*-nitroacetanilide.

7. *Recrystallisation of p-nitroacetanilide.* Dissolve the crude product obtained above in about 20 ml of methylated spirit. Warm to get a clear solution. Filter while hot and cool the filtrate in ice. *o*-nitroacetanilide goes in the filtrate while *p*-nitroacetanilide is obtained as colourless crystals on the filter paper. Wash the solid on the filter paper with cold water.
8. Dry the solid, weigh it and record its yield.

RESULT

Weight of *p*-nitroacetanilide is obtained = ... g

Melting point of the compound is ...... °C

Note: *Approximate expected yield is 4 g.*

The melting point of p-nitroacetanilide is 214°C.

PRECAUTIONS

1. During addition of nitrating mixture, the temperature of the reaction mixture should not rise above 10°C.
2. Addition of fuming nitric acid should be done dropwise.
3. Do not inhale the vapours of nitric acid as they are very corrosive in nature. Addition of nitrating mixture may preferrably be done in a fume-cupboard.

EXPERIMENT 8.4

AIM: ***To prepare 2-naphthol aniline dye.***

THEORY

2-Naphthol aniline dye or Phenyl-azo-β-naphthol is an orange-red dye. It belongs to a large class of azo-compounds, all of which contain the characteristic grouping $>C-N=N-C<$. Azo compounds are all coloured compounds. For the preparation of this dye, aniline is diazotised and then diazonium salt thus obtained is subjected to coupling reaction with 2-naphthol.

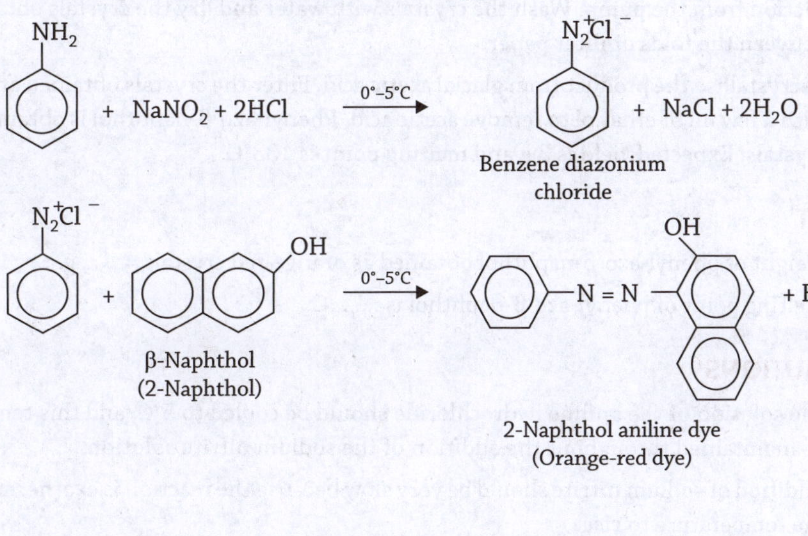

APPARATUS

One 100 ml conical flask, one 100 ml beaker, one 250 ml beaker, ice-bath, glass rod, buchner funnel, water pump.

CHEMICALS

Aniline	= 4.5 ml
Sodium nitrite	= 4 g
2-Naphthol	= 7 g
Conc. hydrochloric acid	= 10 ml
Glacial acetic acid	= 40 ml

PROCEDURE

1. Take a 100 ml conical flask and add 4.5 ml of aniline, 10 ml of conc. HCl and 20 ml of water. Cool this solution to 5°C by placing the conical flask in a trough containing icecold water.
2. In a 100 ml beaker dissolve 4 g of sodium nitrite in 20 ml of water and cool this solution also to 5°C.
3. Now slowly add sodium nitrite solution to the solution of aniline in conc. HCl.
4. Dissolve 7.0 g of 2-naphthol in 60 ml of 10% NaOH solution taken in a 250 ml beaker and cool this solution to 5°C by placing in an ice bath. Some crushed ice may be added directly to fecilitate cooling.
5. Now add the diazotised solution *very slowly* to the 2-naphthol solution with constant stirring. The mixed solutions immediately develop a red colour and 2-naphthol aniline rapidly separates as orange-red crystals.
6. When the addition of diazo solution is complete, allow the mixture to stand in ice-salt mixture for 30 minutes, with occasional stirring. Filter the solution through a buchner funnel under suction from the pump. Wash the crystals with water and dry the crystals obtained by pressing between the folds of filter paper.
7. Recrystallise the product from glacial acetic acid. Filter the crystals obtained at the pump. Wash with a few ml of ethanol to remove acetic acid. Phenyl-azo-β-naphthol is obtained as orange-red crystals. Expected yield is 3 g and melting point is 133°C.

RESULT

Weight of phenyl-azo-β-naphthol obtained as orange-red crystals = ... g.

Melting point of phenyl-azo-β-naphthol is ...... °C.

PRECAUTIONS

1. The solution of the aniline hydrochloride should be cooled to 5°C, and this temperature should be maintained throughout the addition of the sodium nitrite solution.
2. Addition of sodium nitrite should be very slow because the reaction is exothermic and may cause the temperature to rise.
3. Always add diazonium chloride solution to β-naphthol solution for dye formation and not vice versa.

VIVA QUESTIONS WITH ANSWERS

Q. 1. What is Claisen-Schmidt reaction ?

Ans. It is the condensation of aromatic aldehydes with aliphatic ketones (or aryl alkyl ketones) in the presence of dil. NaOH solution.

Q. 2. What is IUPAC name of acetanilide ?

Ans. Phenylethanamide.

Q. 3. What is the function of zinc dust in the preparation of acetanilide ?

Ans. (*i*) It reduces the coloured impurities present in aniline.

(*ii*) It prevents oxidation of aniline during the reaction.

Q. 4. Give the formula of 2-naphthol aniline dye.

Ans.

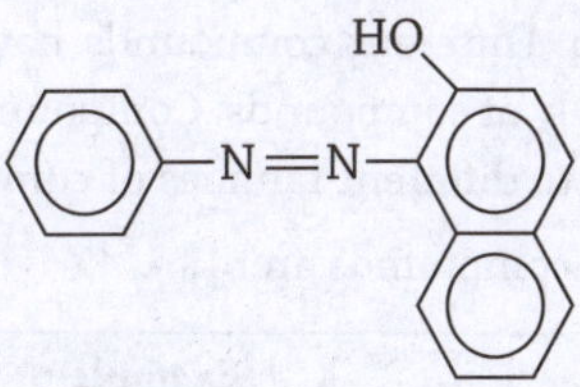

Q. 5. What is the colour of 2-naphthol aniline dye ?

Ans. Orange-red.

Q. 6. What is diazotisation reaction ?

Ans. It is the reaction of primary aromatic amines with nitrous acid to form diazonium salt. The reaction is carried out at low temperature (below 5°C)

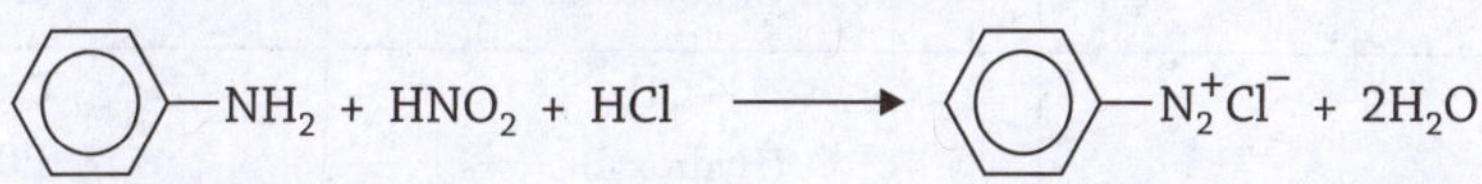

Q. 7. What is coupling reaction ?

Ans. It is the reaction of diazonium salts with highly activated benzene rings such as phenolic ring. The reaction involves electrophilic substitution and the product obtained is generally a dye.

Q. 8. Which of the following compounds on diazotisation followed by compling with β-naphthol will form an azo dye ?

(*i*) *p*-Toluidine **(*ii*) Benzylamine**

(*iii*) N-Methylaniline

Ans. *p*-Toluidine, because it is a primary aromatic amine.

CHAPTER 9

Tests for the Functional Groups Present in Organic Compounds

Molecules of organic compounds except that of hydrocarbons can be divided into two parts, *a reactive part* which is referred to as **functional group** and a *skeleton of carbon atoms* called **alkyl group.** The properties of a compound are largely determined by the functional group. Different compounds having same functional group have similar properties and are classified as family of compounds. Compounds having different functional groups have different properties and belong to different families of compounds.

Some of the common functional groups present in organic compounds are:

Class of compounds	***Functional group***	***Example***
Olefins (alkenes)	$>C=C<$ (Carbon-carbon double bond)	$CH_2=CH_2$ (Ethene)
Acetylenes (alkynes)	$-C\equiv C-$ (Carbon-carbon triple bond)	$CH\equiv CH$ (Ethyne)
Alcohols	$-OH$ (Hydroxyl)	C_2H_5OH (Ethanol)
Carboxylic acids	$-\overset{\overset{\large O}{\Vert}}{C}-OH$ (Carboxyl)	CH_3COOH (Ethanoic acid)
Aldehydes	$-\overset{\overset{\large H}{\vert}}{C}=O$ (Aldehydic)	CH_3CHO (Ethanal)
Ketones	$>C=O$ (Ketonic)	$(CH_3)_2C=O$ (Propanone)
Amines	$-NH_2$ (Amino)	CH_3NH_2 (Methanamine)

9.01. HYDROCARBONS

Compounds containing only carbon and hydrogen are called hydrocarbons. The other organic compounds are derived from hydrocarbons by replacement of one or more hydrogen atoms by other atoms or group of atoms (functional group) such as —OH, —CHO, —COOH, —Cl, etc. The hydrocarbons are classified into saturated and unsaturated hydrocarbons.

SATURATED HYDROCARBONS (Alkanes)

These hydrocarbons contain single bonds only between C—C and between C—H. These are saturated hydrocarbons because the four valencies of all carbon atoms are satisfied with single bonds. Because of the low reactivity, they are also called **paraffins.** Since carbon atoms can form long chains—straight as well as branched and rings, they are divided into two types: *alkanes or aliphatic hydrocarbons* and *cycloalkanes*. Some of the members of alkanes are:

CH_4	$CH_3CH_2CH_3$	$CH_3CH_2CH_2CH_3$	$CH_3CH_2CH_2CH_2CH_2CH_3$
Methane	Propane	Butane	Hexane

UNSATURATED HYDROCARBONS (Alkenes and Alkynes)

These are the hydrocarbons which contain multiple (double or triple) carbon-carbon bonds.

The aliphatic hydrocarbons that contain a double bond ($-\overset{|}{C}=\overset{|}{C}-$) between two carbon atoms are called **alkenes.** Some of its members are:

$CH_2=CH_2$	$CH_3CH=CH_2$	$CH_3.CH_2.CH=CH_2$
Ethene	Propene	But-1-ene

A σ bond and a π bond constitute a double bond.

The aliphatic hydrocarbons that contain a triple bond ($-C\equiv C-$) between two carbon atoms are called **alkynes.** Some of its members are:

$CH\equiv CH$	$CH_3.C\equiv CH$	$CH_3.CH_2C\equiv CH$
Ethyne	Propyne	But-1-yne

A σ bond and two π bonds constitute a triple bond.

The number of π bonds present in a molecule of an organic compound is termed as **degree of unsaturation.**

TESTS FOR UNSATURATION

There are two tests for detecting unsaturation in organic compound:

1. *Bromine Test.*
2. *Alkaline $KMnO_4$ Test.*

1. BROMINE TEST

In this test, the orange-red colour of bromine solution disappears when it is added to the unsaturated hydrocarbon to form colourless addition products.

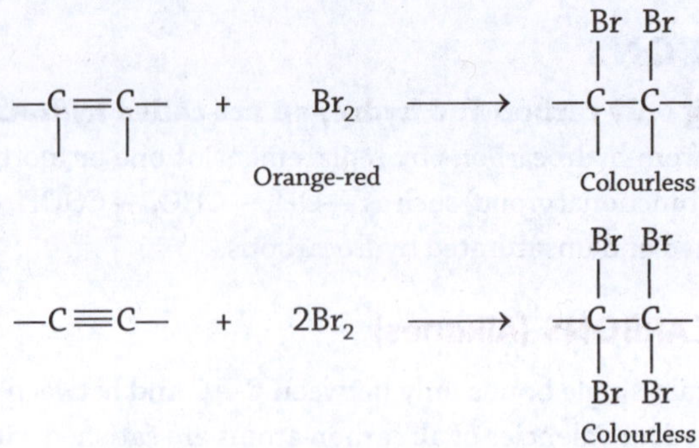

Procedure

(i) When the compound is soluble in water. Dissolve a part of the given compound (solid or liquid) in about 2 ml of distilled water in a test-tube and add a drop of bromine water and shake.

Disappearance of orange-red colour of bromine indicates unsaturation. Continue the addition of bromine water dropwise with constant shaking. The disappearance of orange-red colour continues so long as there are unsaturated bonds. When all the pi-bonds are broken, the orange-red colour shall persist. Bring a rod dipped in NH_4OH near the mouth of the test-tube.

Absence of white fumes confirms unsaturation.

(ii) When the compound is insoluble in water. Dissolve a small amount of the given compound in 1 ml of CCl_4 in a test tube and add 2% solution of bromine in CCl_4. Shake the mixture.

Disappearance of orange-red colour indicates unsaturation. Continue the dropwise addition of Br_2 with constant shaking until the brown colour persists. Bring a rod dipped in NH_4OH near the mouth of the test-tube.

Absence of white fumes confirms unsaturation.

(iii) When the compound is a gas. Add 1–2 ml of bromine solution into a gaseous compound taken in a jar and shake. Disappearance of orange-red colour of bromine indicates unsaturation.

Note: *No HBr is evolved in bromine test for unsaturation. However, some compounds such as aniline form substituted products with bromine and evolve HBr.*

> **CAUTION !**
> ***Handle bromine solution carefully. Avoid its contact with skin.***

2. ALKALINE $KMnO_4$ TEST (BAEYER'S TEST)

In this test, the pink colour of $KMnO_4$ disappears, when an alkaline $KMnO_4$ is added to an unsaturated hydrocarbon. The disappearance of pink colour may take place with or without the formation of brown precipitate of MnO_2.

$$2KMnO_4 + H_2O \longrightarrow 2KOH + 2MnO_2 + 3[O]$$

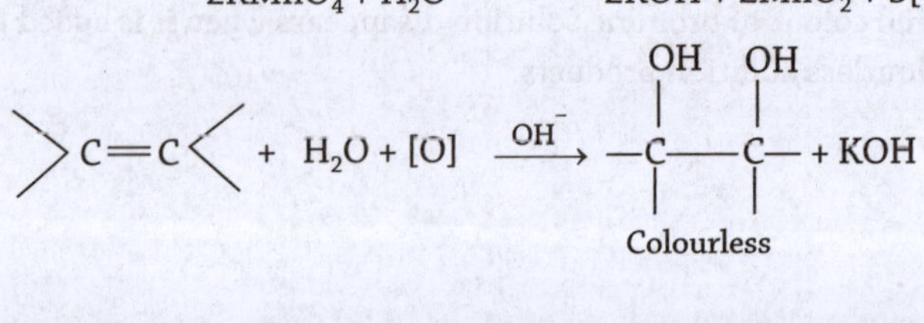

Procedure

Dissolve a little of the given organic compound in about 2 ml of water or acetone in a test tube. Add 1–2 drops of alkaline solution of $KMnO_4$ (1%) and shake the mixture.

Decolourization of pink colour of $KMnO_4$ indicates unsaturation.

PREPARATION OF REAGENTS

1. **Bromine water.** Add 2 ml of liquid bromine in 100 ml of distilled water and shake. Decant off the clear liquid.
2. **Bromine in CCl_4.** Shake 2 ml of liquid bromine in 100 ml of CCl_4 and stopper the bottle.
3. **Alkaline $KMnO_4$ (1%).** Dissolve 1 g of $KMnO_4$ (solid) in 100 ml of distilled water. Now add 10 g of anhydrous Na_2CO_3. Shake to dissolve and stopper the bottle.

9.02. ALCOHOLS

Compounds in which the hydroxyl group (—OH) is linked to aliphatic carbon chain or in the side chain of an organic compound are called **alcohols.** The alcohols containing one, two or three hydroxyl groups per molecule are called mono, di or trihydric alcohols respectively.

Alcohols are further classified as **primary (1°), secondary (2°) and tertiary (3°)** according as the —OH group is attached to primary, secondary or tertiary carbon atoms respectively.

CH_3CH_2OH
Ethyl alcohol
(Primary)

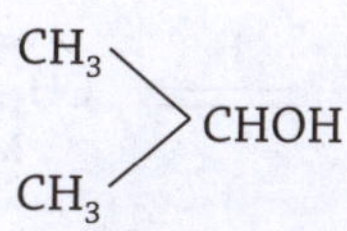

iso-Propyl alcohol
(Secondary)

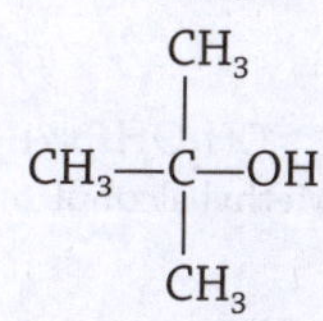

tert-Butyl alcohol
(Tertiary)

TESTS FOR THE ALCOHOLIC [R—OH] GROUP

The alcoholic group can be detected by any of the following tests:

1. *Sodium metal test.*
2. *Ester test.*
3. *Ceric ammonium nitrate test.*
4. *Iodoform test for alcohols containing the* $CH_3-CH(OH)-$ *group*

1. SODIUM METAL TEST

This test is based on the appearance of effervescence due to liberation of hydrogen gas when the alcohol is reacted with active metals like sodium.

$$\underset{\text{Alcohol}}{2R-OH} + 2Na \longrightarrow \underset{\text{Sodium alkoxide}}{2R-O^-Na^+} + H_2\uparrow$$

$$\underset{\text{Methanol}}{2CH_3OH} + 2Na \longrightarrow \underset{\text{Sodium methoxide}}{2CH_3-O^-Na^+} + H_2\uparrow$$

Procedure

Take about 1 ml of the given pure liquid in a dry test tube, add about 1 gram of anhydrous calcium sulphate and shake well to remove water. Filter or decant off the liquid to another clean dry test tube and add a small piece of sodium metal.

A brisk effervescence due to the evolution of hydrogen gas indicates an alcoholic group.

Notes: 1. *The alcohol should be dry as water also reacts with sodium metal to evolve hydrogen gas.*

2. *Do not throw unreacted sodium metal into the sink or waste bin. Add more alcohol to the unreacted sodium to complete the reaction.*

2. ESTER TEST

Alcohols react with carboxylic acids to form sweet smelling esters. The reaction between an alcoholic group and a carboxylic acid group is called *esterification*. This is a slow and reversible reaction and is catalysed by an acid such as conc. sulphuric acid.

$$\underset{\text{Alcohol}}{R—OH} + \underset{\text{Carboxylic acid}}{R'—\overset{\overset{\large O}{\|}}{C}—OH} \xrightleftharpoons{H^+} \underset{\text{Ester}}{R'—\overset{\overset{\large O}{\|}}{C}—O—R} + H_2O$$

$$\underset{\text{Methyl alcohol}}{CH_3OH} + \underset{\text{Acetic acid}}{CH_3—\overset{\overset{\large O}{\|}}{C}—OH} \xrightleftharpoons{H^+} \underset{\text{Methyl acetate}}{CH_3—\overset{\overset{\large O}{\|}}{C}—O—CH_3} + H_2O$$

Procedure

Take 1 ml of the given liquid in a clean dry test tube, add 1 ml of glacial acetic acid and 2–3 drops of conc. sulphuric acid. Warm the mixture on a water bath for about 10 minutes. Pour it into about 20 ml of cold water taken in a beaker and smell.

A fruity smell indicates the presence of an alcoholic group.

3. CERIC AMMONIUM NITRATE TEST

Place about 1 ml of the given compound in a clean dry test-tube, add a few drops of ceric ammonium nitrate reagent and shake well.

Appearance of pink or red colour indicates the presence of an alcoholic group.

$$\underset{\text{Alcohol}}{2R—OH} + \underset{\text{Ceric ammonium nitrate}}{(NH_4)_2Ce(NO_3)_6} \longrightarrow \underset{\text{Pink or red complex}}{[(ROH)_2Ce(NO_3)_4]} + 2NH_4NO_3$$

$$2CH_3—OH + (NH_4)_2Ce(NO_3)_6 \longrightarrow \underset{\text{Pink or red complex}}{[(CH_3OH)_2Ce(NO_3)_4]} + 2NH_4NO_3$$

4. IODOFORM TEST

This test is given by acetaldehyde, all methyl ketones and all alcohols containing $CH_3—\underset{\underset{OH}{|}}{CH}—$ group. The compound is warmed with sodium hydroxide solution and iodine. A yellow ppt. of iodoform is obtained.

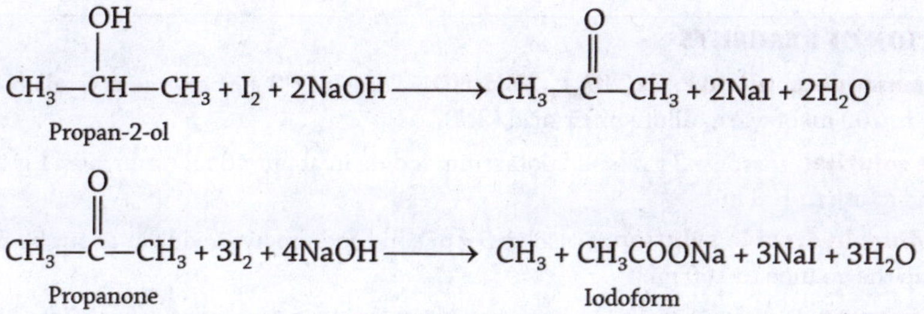

Procedure

Take 1 ml of the given compound in a clean dry test tube and add about 1 ml of 1% iodine solution. Then add dilute sodium hydroxide solution dropwise until the brown colour of iodine is discharged. Warm gently on a water bath.

Formation of yellow precipitate of iodoform indicates ethanol, ethanal, sec-methyl alcohol or methyl ketone.

DISTINCTION BETWEEN PRIMARY, SECONDARY AND TERTIARY ALCOHOLS

LUCAS TEST

This test is based upon relative reactivities of various alcohols towards HCl in the presence of $ZnCl_2$. In this test, the alcohol is treated with Lucas reagent which is equimolar mixture of HCl and $ZnCl_2$. Alcohols are soluble in Lucas reagent, so a clear solution is obtained. On reaction, alkyl chlorides are formed which being insoluble result in cloudiness in the solution.

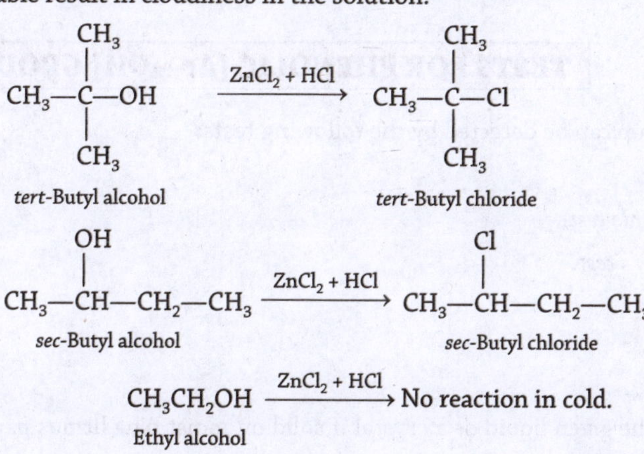

Procedure

Take about 1 ml of dry alcohol in a clean dry test tube and add 8–10 drops of Lucas reagent. Shake the mixture well. Note down the appearance of cloudiness.

(*i*) *If cloudiness appears immediately,* **tertiary alcohol** *is indicated.*

(*ii*) *If cloudiness appears within five minutes,* **secondary alcohol** *is indicated.*

(*iii*) *If cloudiness appears only upon heating,* **primary alcohol** *is indicated.*

PREPARATION OF REAGENTS

1. **Ceric ammonium nitrate. $Ce(NO_3)_4.2NH_4NO_3$.** Dissolve 20 g of orange crystals of ceric ammonium nitrate in 200 ml of warm dilute nitric acid (2M).
2. **Iodine solution.** Dissolve 2 g of solid potassium iodide in about 40 ml water, add 1 g solid iodine, shake well and dilute to 100 ml.
3. **5% sodium hydroxide solution.** Dissolve 5 g of solid sodium hydroxide in about 50 ml water and then make up the volume to 100 ml.
4. **Lucas reagent.** Dissolve 135 g of anhydrous zinc chloride in 100 ml of concentrated hydrochloric acid. Shake well and cool.

9.03. PHENOLS

Compounds containing one or more hydroxyl groups attached to an aromatic ring are called phenols. C_6H_5OH is the simplest phenol. Other examples are: *o*, *m* and *p*-cresol, catechol and resorcinol. All except *m*-cresol are solids, *m*-cresol is a liquid. Phenol C_6H_5OH is a solid in winter and liquid in summer (*m.p.* = 40.9°C). All are colourless when pure but generally slightly coloured due to oxidation. They have low solubility in water but have appreciable solubility in alcohol.

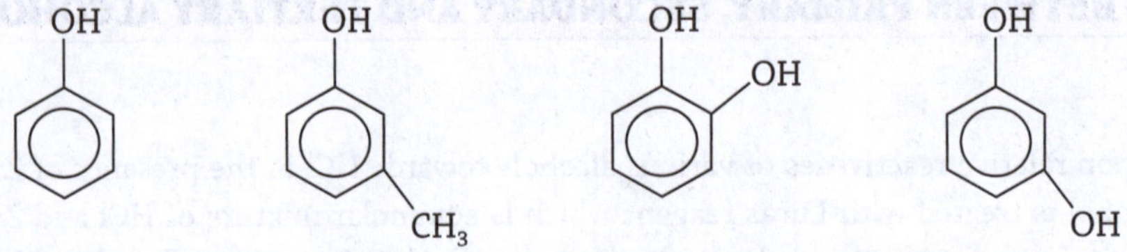

Phenol *m*-Cresol Catechol Resorcinol

TESTS FOR PHENOLIC [Ar—OH] GROUP

The phenolic group can be detected by the following tests:

1. *Litmus test.*
2. *Ferric chloride test.*
3. *Liebermann's test.*

1. LITMUS TEST

Procedure

Place a drop of the given liquid or a crystal if solid on moist blue litmus paper. *If the colour changes to red, phenolic group may be present.*

Note: Carboxylic acids also give this test. But as phenol is not as strong an acid as carboxylic acid, it does not give an effervescence with aqueous solution of sodium carbonate.

2. FERRIC CHLORIDE TEST

Phenols react with ferric ions to form coloured complexes. Many other types of organic compounds also react with ferric chloride solution.

Procedure

Take 1 ml of neutral ferric chloride solution in a clean test tube and 2–3 drops of the liquid compound (or 1 crystal if solid). A change in colour indicates the phenolic group. *Phenol produces violet colouration.*

$$\underset{\text{Ferric chloride}}{FeCl_3} + \underset{\text{Phenol}}{6C_6H_5OH} \longrightarrow \underset{\text{Violet complex}}{[Fe(OC_6H_5)_6]^{3-}} + 3HCl$$

Notes:

1. *o, m, p-cresol, resorcinol give violet or blue colouration.*
2. *β-Naphthol gives a green colouration.*
3. *α-Naphthol gives pink colouration.*
4. *Formic acid and acetic acid give deep red colouration.*

Preparation of neutral ferric chloride. Place 1 ml of 1% ferric chloride solution in a clean test-tube and add ammonia solution until a brown precipitate just appears. Now add the original ferric chloride solution dropwise until the precipitate just disappears.

3. LIEBERMANN'S TEST

Most of the phenols give this test. Red or brown colouration is produced when a nitrite is added to a phenol dissolved in conc. sulphuric acid. The colour changes to blue or green by the addition of a strong alkali.

$$\underset{\text{Sodium nitrite}}{2NaNO_2} + H_2SO_4 \longrightarrow \underset{\text{Nitrous acid}}{2HNO_2} + Na_2SO_4$$

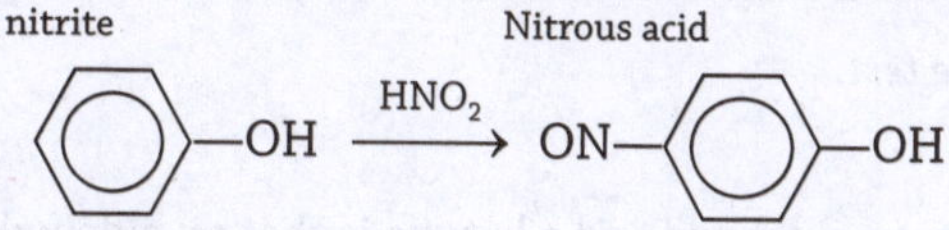

Phenol p-Nitrosophenol

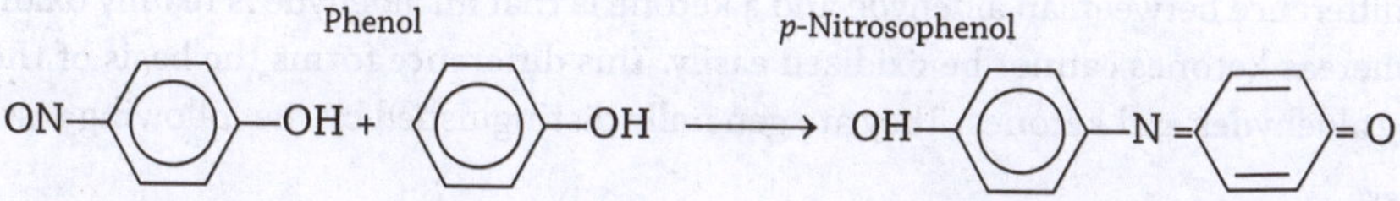

Indophenol (red)

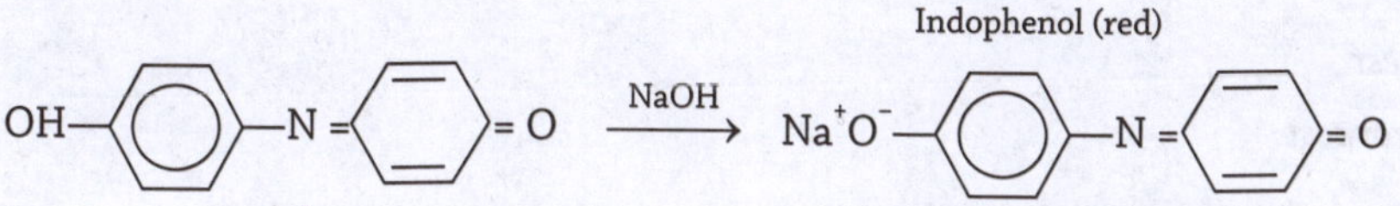

red Indophenol anion (blue)

Procedure

Place 2–3 crystals of sodium nitrite in a clean dry test tube and add about 1 g of phenol. Heat very gently for half a minute and allow it to cool. Then add about 1 ml of conc. sulphuric acid and shake the tube to mix the contents. *A deep blue or deep green colouration develops*. Add water carefully, the colour turns red. Now add an excess of sodium hydroxide solution, the blue or green colour appears.

Notes:

1. *Nitrophenols and p-substituted phenols do not give this test.*
2. *Among the dihydroxyphenols, only resorcinol gives positive test.*

4. PHTHALEIN DYE TEST

Take 0.1 g of organic compound and 0.1 g of phthalic anhydride in a clean dry test tube and add 1–2 drops of conc. H_2SO_4. Heat the test tube for about 1 minute in an oil bath. Cool and pour the reaction mixture carefully into a beaker containing 15 mL of dilute sodium hydroxide solution. **Appearance of pink, blue, green, red etc. colours indicates the presence of phenolic —OH group in the compound.**

9.04. ALDEHYDES AND KETONES

Aldehydes and ketones are the compounds that contain the carbonyl group, $-\overset{O}{\overset{\|}{C}}-$. This group has characteristic properties which are shown by both classes of compounds. In aldehydes, the carbonyl group is attached to a hydrogen atom and to an aliphatic or aromatic radical. Formaldehyde is an exception in which the carbonyl group is attached to two hydrogen atoms.

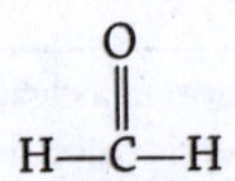
Formaldehyde

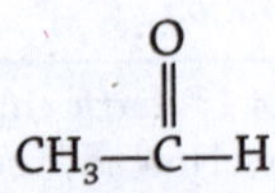
Acetaldehyde

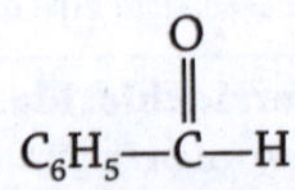
Benzaldehyde

In ketones, the carbonyl group is attached to two aliphatic or aromatic radicals.

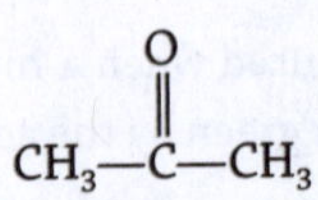
Acetone

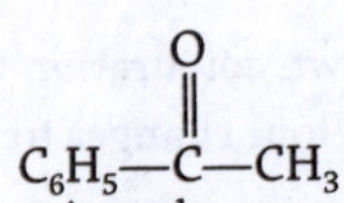
Acetophenone

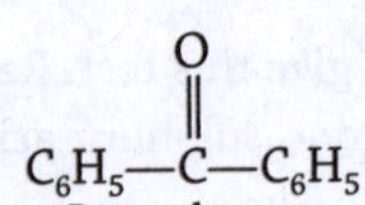
Benzophenone

The carbonyl group in aldehydes and ketones is identified by the following tests:

1. *2, 4-dinitrophenyl hydrazine test.*
2. *Sodium bisulphite test.*

The major difference between an aldehyde and a ketone is that an aldehyde is readily oxidized to carboxylic acid whereas ketones cannot be oxidized easily. This difference forms the basis of the tests for distinguishing aldehydes and ketones. They are generally distinguished by the following tests:

1. *Schiff's test*
2. *Fehling's test*
3. *Silver mirror test.*

TESTS FOR ALDEHYDIC AND KETONIC [—CHO and —CO—] GROUPS

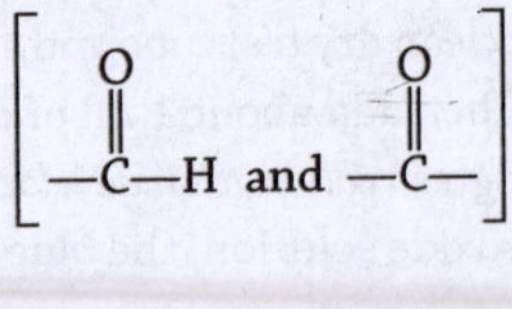

1. 2, 4-DINITROPHENYL HYDRAZINE TEST (2, 4—DNP TEST)

Take 0.5 ml or 0.5 g of the given compound in a clean dry test-tube, add rectified spirit until the compound just dissolves. Now add a few drops of 2, 4-dinitrophenyl hydrazine solution. Cork the test-tube, shake the mixture and allow it to stand for 5 minutes.

Formation of yellow or orange crystals indicate the presence of carbonyl group.

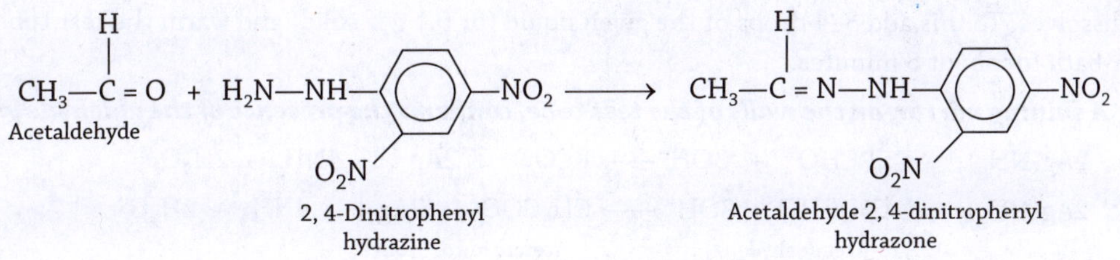

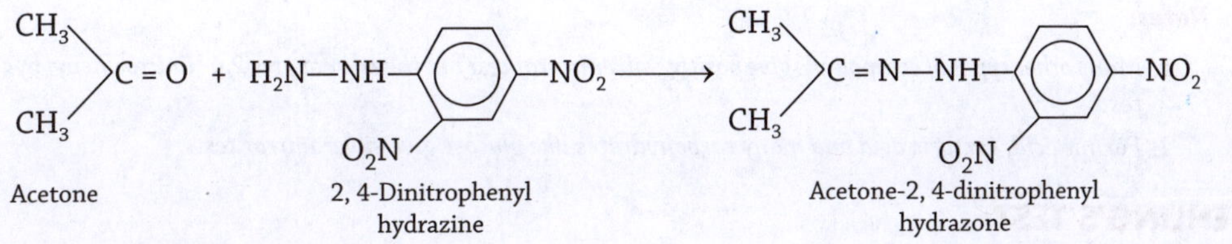

2. SODIUM BISULPHITE TEST

Take 2 ml of a saturated solution of sodium bisulphite in a clean test-tube, and add 1 ml or 1 g of the given compound. Cork the test-tube, shake and leave it for 15–20 minutes.

Formation of crystalline precipitate confirms carbonyl group.

Notes:

1. *Formaldehyde, acetaldehyde, benzaldehyde, acetone, methyl ethyl ketone and diethyl ketone give positive tests.*
2. *Aqueous solutions of formaldehyde and acetaldehyde form addition products but as they are highly soluble, precipitates are rarely formed.*
3. *Acetophenone and benzophenone do not give this test.*

DIFFERENTIATING TESTS FOR ALDEHYDES

The following tests are given by aldehydes but not by ketones:

1. SCHIFF'S TEST

Dissolve about 0.5 ml or 0.5 g of the given compound in alcohol in a clean test-tube and add 1 ml of Schiff's regent. Shake and note the change in colour.

Appearance of pink, red or magenta colour confirms the presence of aldehydic group.

Notes:

1. *The Schiff's reagent should not be warmed.*
2. *The Schiff's reagent should not be treated with alkalies, otherwise the pink colour develops even in the absence of aldehydes.*
3. *With benzaldehyde the pink colour develops slowly.*

2. TOLLEN'S TEST (SILVER MIRROR TEST)

Procedure

Place 1 ml of silver nitrate solution in a clean test tube and add 2–3 ml of dilute NaOH solution. A brown precipitate forms. Now add dilute ammonia solution dropwise until the brown precipitate of silver oxide just dissolves. To this add 3–4 drops of the given liquid (or 0.1 g if solid) and warm the test tube on a water-bath for about 5 minutes.

A shining mirror, on the walls of the test tube, confirms the presence of the aldehyde group.

$$2Ag(NH_3)_2^+ + RCHO + 3OH^- \longrightarrow RCOO^- + 2Ag\downarrow + 4NH_3 + 2H_2O$$

$$2Ag(NH_3)_2^+ + \underset{\text{Acetaldehyde}}{CH_3CHO} + 3OH^- \longrightarrow \underset{\text{Acetate ion}}{CH_3COO^-} + \underset{\text{Silver}}{2Ag\downarrow} + 4NH_3 + 2H_2O$$

Notes:

1. *Many other types of compounds give positive silver mirror test but they do not give 2,4-dinitro phenyl hydrazine test.*
2. *Formic acid, tartaric acid and many carbohydrates like glucose give silver mirror test.*

3. FEHLING'S TEST

Procedure

Take 1 ml each of Fehling's solution A and B in a test tube. Add 4–5 drops (or 0.2 g) of the given organic compound and warm the test tube in hot water bath for 4–5 minutes.

Appearance of a red precipitate confirms the presence of the aldehydic group.

$$\underset{\text{Aldehyde}}{RCHO} + \underbrace{2Cu^{2+} + 5OH^-}_{\text{Fehling's solution}} \longrightarrow \underset{\text{(Red)}}{Cu_2O(s)}\downarrow + RCOO^- + 3H_2O$$

Notes:

1. *Benzaldehyde may or may not give this test as the reaction is very slow.*
2. *Formic acid also gives this test.*

DIFFERENTIATING TESTS FOR KETONES

The following tests are given by ketones but not by aldehydes:

1. m-DINITROBENZENE TEST

Place 0.5 ml of the given liquid (or 0.5 g of solid) in a clean test tube and add about 0.1 g of finely powdered *m*-dinitrobenzene. Now add about 1 ml of dilute sodium hydroxide solution and shake.

Appearance of violet colour which slowly fades confirms ketonic group.

Note: *Benzophenone does not give this test.*

2. SODIUM NITROPRUSSIDE TEST

The anion of the ketone formed by an alkali reacts with nitroprusside ion to form coloured ion.

$$CH_3COCH_3 + OH^- \longrightarrow CH_3COCH_2^- + H_2O$$

$$\underset{\text{Nitroprusside ion}}{[Fe(CN)_5NO]^{2-}} + CH_3COCH_2^- \longrightarrow \underset{\text{Red colouration}}{[Fe(CN)_5NO.\ CH_3COCH_2]^{3-}}$$

Procedure

Dissolve a crystal of sodium nitroprusside in about 1 ml of distilled water in a clean test tube and then add 0.5 ml (or 0.5 g) of the given compound. Shake and add sodium hydroxide solution dropwise.

A red colouration indicates the ketonic group.

Notes:

1. *Benzaldehyde also gives red colour.*
2. *Benzophenone does not give this test.*

PREPARATION OF REAGENTS

1. **2, 4-Dinitrophenylhydrazine (2, 4-DNP).** Dissolve 1 g of 2, 4-dinitrophenyl hydrazine in 50 ml methanol to which 2 ml of concentrated sulphuric acid is added. Filter, if necessary.
2. **Sodium bisulphite, $NaHSO_3$.** Dissolve about 30 g sodium bisulphite in 100 ml of distilled water.
3. **Schiff's reagent.** Dissolve 0.1 g *p*-rosanaline hydrochloride in 100 ml water and pass sulphur dioxide gas until its red colour is discharged. Filter and use the filtrate.
4. **Fehling's solution A.** Dissolve 35 g of crystalline copper sulphate in 500 ml water and add 2 ml conc. H_2SO_4.
5. **Fehling's solution B.** Dissolve 173 g of Rochelle salt (sodium potassium tartrate) and 60 g sodium hydroxide in 500 ml water.

9.05. CARBOXYLIC ACIDS

The organic compounds containing the carboxylic group, $\text{—}\overset{\overset{\text{O}}{\|}}{\text{C}}\text{—OH}$ are called carboxylic acids.

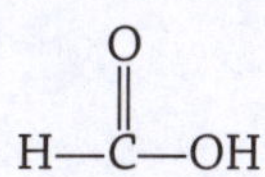

Formic acid

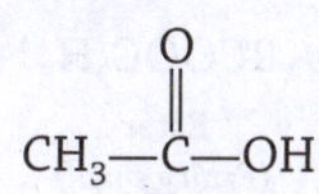

Acetic acid

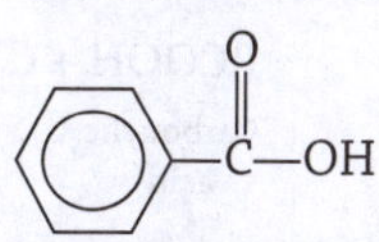

Benzoic acid

Formic acid and acetic acid are liquids. Benzoic acid, oxalic acid, phthalic acid, salicylic acid and tartaric acids are colourless crystalline solids. Aliphatic acids are soluble in water whereas aromatic acids are sparingly soluble in water.

TESTS FOR CARBOXYLIC GROUP [—COOH]

Carboxylic acids can be identified by the following tests:

1. *Litmus test.*
2. *Sodium hydrogencarbonate test.*
3. *Ester test.*

1. LITMUS TEST

The carbobxylic acids turn blue litmus red. The hydroxyl group in —COOH is far more acidic than in alcohol.

$$\text{R—}\overset{\overset{\text{O}}{\|}}{\text{C}}\text{—OH} + H_2O \rightleftharpoons \text{R—}\overset{\overset{\text{O}}{\|}}{\text{C}}\text{—O}^- + H_3O^+$$

Procedure

Place a drop of the given liquid (or a crystal of the solid) on a moist blue litmus paper and note the change in colour.

If the colour changes to red, carboxylic group or phenolic group is present.

Note: *Phenols also give this test.*

2. SODIUM HYDROGENCARBONATE TEST

Carboxylic acids react with sodium hydrogencarbonate to give carbon dioxide gas which is identified by the effervescence produced. This test is used to distinguish carboxylic acids from phenols.

$$\underset{\text{Carboxylic acid}}{RCOOH} + NaHCO_3 \longrightarrow RCOONa + \underset{\text{(Effervescence)}}{CO_2\uparrow} + H_2O$$

Procedure

To 1 ml of organic liquid in a test tube, add a pinch of sodium hydrogencarbonate ($NaHCO_3$).

A brisk effervescence indicates the presence of carboxylic group in the organic compound.

3. ESTER TEST

A carboxylic acid reacts with an alcohol in presence of a little sulphuric acid to form an ester which is recognized by its fruity smell.

$$\underset{\text{Carboxylic acid}}{RCOOH} + C_2H_5OH \xrightarrow{H_2SO_4} \underset{\substack{\text{Ester}\\ \text{(Fruity smell)}}}{RCOOC_2H_5} + H_2O$$

Procedure

Take about 0.1 g of the organic compound, 1 ml of ethyl alcohol and 1–2 drops of conc. H_2SO_4 in a test tube. Heat the reaction mixture on a water bath for about five minutes. Pour the reaction mixture in a beaker containing water.

A fruity smell indicates the presence of carboxylic group in the compound.

9.06. AMINES

Amines may be considered as substitution products of ammonia. When one hydrogen atom of ammonia is replaced with an alkyl or aryl group,the resulting amine is called **primary amine** (R—NH_2). When two hydrogen atoms of ammonia molecule are replaced with two alkyl or aryl groups, the resulting compound is called **secondary amine.** The replacement of all the three hydrogen atoms of ammonia with alkyl and aryl groups gives **tertiary amine.**

Primary amines

CH_3NH_2
Methyl amine

$CH_3CH_2NH_2$
Ethyl amine

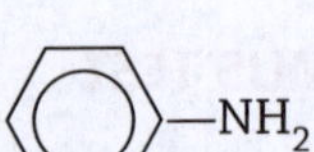
Aniline

Secondary amines

CH_3NHCH_3
Dimethyl amine

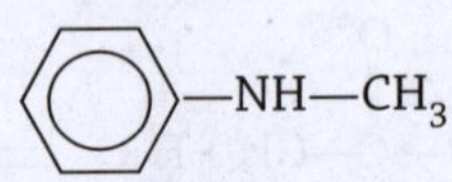
N-Methyl aniline

Tertiary amines

$(CH_3)_3N$
Trimethyl amine

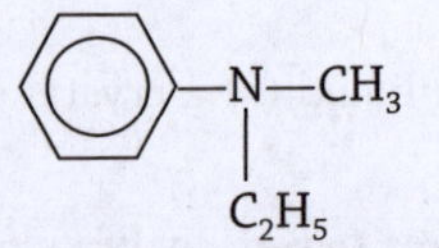
N-Ethyl-N-methyl aniline

TESTS FOR AMINO GROUP [—NH_2]

1. SOLUBILITY TEST

Amines are the organic compounds that have appreciable basicity and so they dissolve in mineral acids.

$$\underset{\text{Aniline}}{C_6H_5NH_2} + HCl \longrightarrow \underset{\text{Anilinium chloride}}{C_6H_5NH_3^+\,Cl^-}$$

Procedure

Take a small amount of the given compound in a clean test-tube and add 2–3 ml of dilute hydrochloric acid. Shake and note the solubility.

If the compound is soluble, it may be an amine.

2. LITMUS TEST

Amines have appreciable basicity and turn red litmus blue.

$$CH_3CH_2NH_2 + H_2O \rightleftharpoons CH_3CH_2NH_3^+ + OH^-$$

Procedure

Place a drop of the given liquid (or pinch of the solid) on the moist red litmus paper and note the change in colour.

If the colour changes to blue, it may be an amine.

3. CARBYLAMINE

This test is given by both aliphatic and aromatic primary amines.

Secondary and tertiary amines do not give this test.

Procedure

Take about 0.2 g of solid potassium hydroxide in a clean dry test tube and add 2 ml of ethanol. Warm the test tube until the pallets dissolve. To this add a few drops of chloroform, small amount of the given compound and warm gently.

An unpleasent odour confirms the presence of primary amine.

$$RNH_2 + CHCl_3 + 3KOH \longrightarrow \underset{\substack{\text{Carbylamine} \\ \text{(Isocyanide)} \\ \text{(Offensive smelling)}}}{RNC} + 3KCl + 3H_2O$$

CAUTION !

1. *Carbylamine (isocyanide) is highly poisonous so do not inhale its vapours.*
2. *Destroy carbylamine with conc. HCl after the test and then throw into sink.*

4. NITROUS ACID TEST

Primary aliphatic amines react with nitrous acid to give nitrogen gas which is seen as bubbles.

$$\underset{\text{Primary amine}}{RNH_2} + HONO \longrightarrow \underset{\text{Alcohol}}{ROH} + H_2O + N_2$$

Secondary amines react with nitrous acid to form a yellow oily nitrosoamine

$$\underset{\text{Secondary-amine}}{R_2NH} + HONO \longrightarrow \underset{\substack{\text{Nitrosoamine} \\ \text{(yellow oil)}}}{R_2N\text{—}NO} + H_2O$$

Tertiary amines react with nitrous acid to form soluble nitrite salts

$$\underset{\text{Tertiary-amine}}{R_3N} + HONO \longrightarrow \underset{\text{Trialkyl ammonium nitrite}}{R_3NH^+ONO^-}$$

Procedure

Make a solution of about 1 g sodium nitrite in about 5 ml water in a test tube and cool in ice bath. In a separate test tube, dissolve a few drops or few crystals of the given compound in about 1 ml conc. hydrochloric acid and cool this also in ice bath. Mix both the solutions and observe that what happens.

(*a*) ***Bubbles of nitrogen gas*** are seen if the compound is a **primary aliphatic amine.**

(*b*) ***A yellow oily layer*** is formed if the compound is a **secondary amine.**

(*c*) ***No visible change*** in the reaction mixture if the compound is a **tertiary amine.**

5. AZO-DYE TEST

This test is given by primary aromatic amines. Primary amines on reaction with nitrous acid give diazonium salts. These diazonium salts can undergo coupling reaction with phenols resulting in the formation of coloured compounds.

$$\underset{\text{Aniline}}{C_6H_5NH_2} + \underset{\text{Nitrous acid}}{HNO_2} + HCl \longrightarrow \underset{\substack{\text{Benzene diazonium} \\ \text{chloride}}}{C_6H_5N^+\equiv NCl^-} + 2H_2O$$

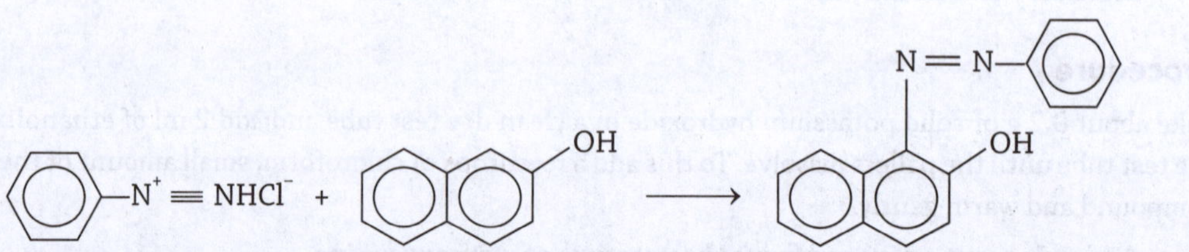

Benzene diazonium chloride β-Napthol Orange-red ppt.

Procedure

Take three test tubes A, B and C. In test tube A, dissolve about 0.2 g of the compound in 2 ml of dilute hydrochloric acid. In test tube B, prepare an aqueous solution of sodium nitrite.

In test tube C, dissolve 0.2 g of β-naphthol in dilute sodium hydroxide.

Place all the three test tubes in an ice bath.

Now add sodium nitrite solution to test tube A and the resulting solution is added to the test tube C.

Formation of a red or orange dye confirms the presence of primary aromatic amino (Ar–NH$_2$) group.

DISTINGUISHING TEST FOR PRIMARY, SECONDARY AND TERTIARY AMINES

1. HINSBERG TEST

The given amine is shaken with benzene sulphonyl chloride.

(*i*) A clear solution in NaOH solution which on acidification gives an insoluble material indicates *primary amine.*

(*ii*) A precipitate (insoluble compound) which is insoluble in NaOH solution indicates *secondary amine.*

(*iii*) Tertiary amines do not react with benzene sulphonyl chloride. An insoluble compound in NaOH solution which dissolves *by the addition of acid indicates tertiary amine.*

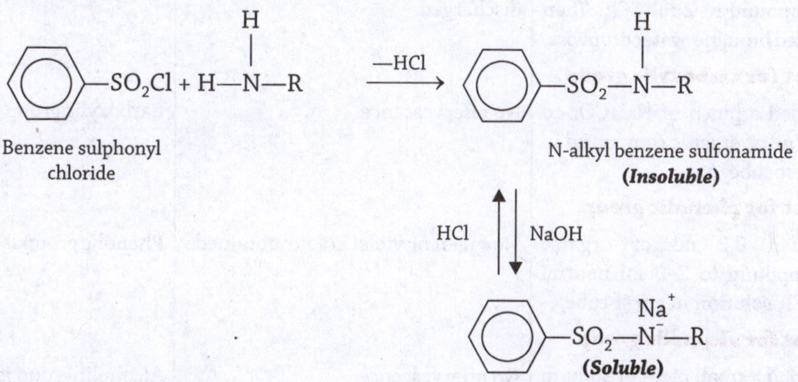

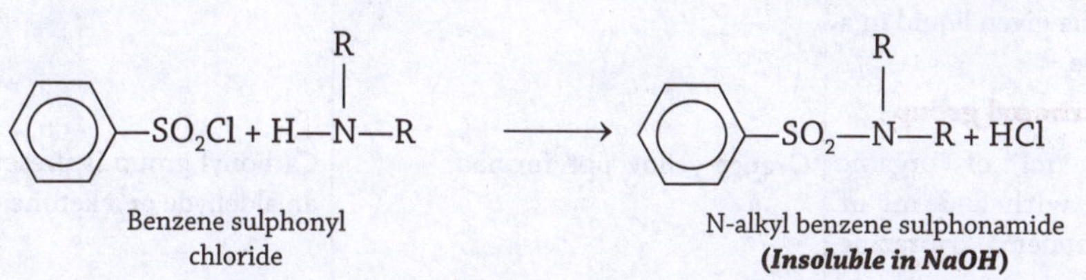

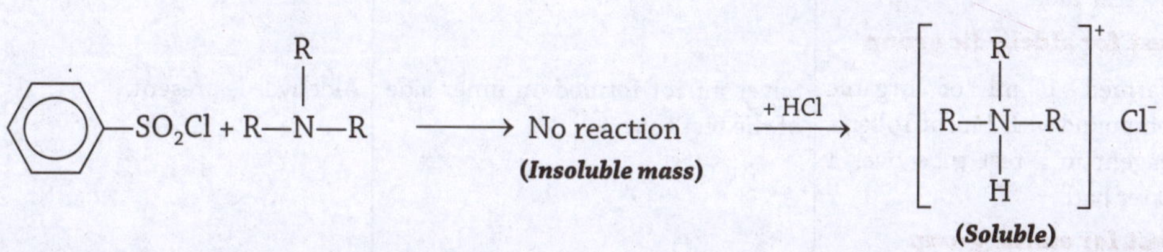

Procedure

Place 0.5 ml of the given amine in a clean test tube, add about 2 ml of 25% NaOH, 2 ml of water and 1 ml of benzene sulphonyl chloride. Shake the mixture for about 10 minutes and then cool under tap water and note the formation of a precipitate. Treat the precipitate, if any, with 2 ml of conc. HCl.

(*i*) Tertiary amine. Precipitate in the test tube. It dissolves in conc. HCl.

(*ii*) Secondary amine. Precipitate in the test tube. It does not dissolve in conc. HCl.

(*iii*) Primary amine. No precipitate (clear solution). On addition of conc. HCl, insoluble material is seen.

SPECIMEN RECORD OF ANALYSIS OF FUNCTIONAL GROUP IN AN ORGANIC COMPOUND

EXPERIMENT 9.1

AIM: ***To identify the functional group present in the given organic compound.***

Experiment	*Observations*	*Inference*
1. Test for unsaturation		
Dissolved 0.2 ml of organic compound in 2 ml CCl_4. Then added bromine-water dropwise.	Brown colour of bromine not discharged.	No unsaturation is present.
2. Test for carboxylic group		
Added a pinch of $NaHCO_3$ to 0.2 ml of organic compound in a test tube.	No effervescence.	Carboxylic group is absent.
3. Test for phenolic group		
Added 0.2 ml of organic compound to 2–3 ml neutral $FeCl_3$ solution in a test tube.	No green or violet colour obtained.	Phenolic group is absent.
4. Test for alcoholic group		
Added a small piece of sodium to 1 ml of the given liquid in a dry test tube.	No effervescence.	Alcoholic group is absent.
5. Test for carbonyl group		
Shook 0.2 ml of organic compound with 2–3 ml of 2, 3-dinitrophenyl hydrazine in a test tube.	Orange-yellow ppt. formed.	Carbonyl group is present. May be an aldehyde or a ketone.
6. Test for aldehydic group		
Warmed 1 ml of organic compound with 1 ml of Tollen's reagent in a test tube over a water bath.	Silver mirror formed on inner side of the test tube.	Aldehyde is present.
7. Test for amine group		
To a small amount of organic liquid in test tube, added 1 ml conc. of HCl and a few drops of $CHCl_3$. Then, added 2 ml of alc. KOH solution and warmed test tube.	No offensive smelling gas evolved.	Amino group absent.

RESULT

The given organic compound contains aldehydic $\left(\begin{array}{c}—\underset{\|}{C}—H\\ O\end{array}\right)$ functional group.

VIVA QUESTIONS WITH ANSWERS

Q. 1. What is a functional group?

Ans. The group of atoms that largely determines the properties of an organic compound is called functional group.

Q. 2. Name any four functional groups.

Ans. Hydroxyl group (—OH)

Amino group (—NH_2)

Carboxyl group (—COOH)

Aldehydic group (—CHO).

Q. 3. Name the functional groups present in alkenes and alkynes.

Ans. Alkenes are unsaturated hydrocarbons with carbon-carbon double bond (C =C) bond present in them. Alkynes are unsaturated hydrocarbons with carbon-carbon triple bond ($C\equiv C$) bond present in them.

Q. 4. What is Baeyer's test for unsaturation?

Ans. When Baeyer's reagent (alkaline potassium permanganate) is added to unsaturated compound, its colour gets discharged indicating presence of C =C or $C\equiv C$ in the compound.

Q. 5. Do alkynes turn blue litmus paper red?

Ans. No.

Q. 6. Which is more acidic—an alcohol or a phenol?

Ans. A phenol is more acidic.

Q. 7. Why is alcohol dried before carrying out sodium metal test?

Ans. Because water also reacts with sodium and gives hydrogen gas with brisk effervescence.

Q. 8. What is the use of Lucas reagent?

Ans. It is used to distinguish between primary, secondary and tertiary alcohols.

Q. 9. Which of the two is more acidic—phenol or carboxylic acid?

Ans. Carboxylic acid is more acitic.

Q. 10. Name a test by which you can distinguish between hexylamine ($C_6H_{13}NH_2$) and aniline ($C_6H_5NH_2$).

Ans. Dye test.

Q. 11. Name two tests which distinguish aldehydes from ketones.

Ans. Tollen's test and Fehling's test.

Q. 12. Name a reagent used to detect carbonyl group in a compound.

Ans. DNP (2, 4-dinitrophenylhydrazine).

Q. 13. What is Tollen's reagent?

Ans. It is ammonical silver nitrate solution.

Q. 14. What is the use of Schiff's reagent?

Ans. Schiff's reagent is used to detect aldehyde group.

Q. 15. What is Rochelle's salt?

Ans. Sodium potassium tartarate is called Rochelle's salt.

Q. 16. What is Fehling's solution?

Ans. It is a solution obtained by mixing equal volumes of copper sulphate solution (Fehling A) and a solution of sodium hydroxide containing sodium potassium tartarate (Fehling B).

Q. 17. How is nitrous acid is prepared?

Ans. When sodium nitrite is reacted with dil. HCl at a temperature below 5°C, nitrous acid is produced.

Q. 18. What is application of carbylamine reaction?

Ans. It is used to detect primary amine.

Q. 19. How can phenol and aniline be distinguished chemically?

Ans. Phenol is soluble in aqueous NaOH solution whereas aniline is not.

Aniline is soluble in dilute HCl whereas phenol is not.

Q. 20. In contrast to aromatic primary amines, aliphatic primary amines do not form stable diazonium salts. Why?

Ans. Because alkyl carbocation formed on decomposition of diazonium salt is more stable than phenyl carbocation.

Q. 21. Why is aniline weaker base than ammonia?

Ans. Because lone pair of nitrogen in aniline is delocalized over benzene ring and is not fully available for sharing with acids.

Q. 22. How can you distinguish between methanol and ethanol chemically?

Ans. Methanol and ethanol can be distinguished by iodoform test. Ethanol gives yellow ppt. of iodoform in this test whereas methanol does not give this test positive.

CHAPTER 10

Tests of Carbohydrates, Fats and Proteins in Pure Samples and Detection of their Presence in Given Food Stuffs

Food is a necessary material which must be supplied to the body for its normal and proper functioning.

The essential constituents of food are:

1. *Carbohydrates* 2. *Lipids* (oils and fats) 3. *Proteins* 4. *Minerals* 5. *Vitamins and* 6. *Water*.

10.01. CARBOHYDRATES

Carbohydrates are polyhydroxy aldehydes, polyhydroxy ketones, their derivatives and the substances which yield them on hydrolysis.

The carbohydrates which are ketones are called **ketoses** and those that are aldehydes are called **aldoses.** The general term for all the carbohydrates is **glycose.**

The carbohydrates which cannot be hydrolysed to simpler carbohydrates are called **monosaccharides.** For example, glucose, fructose, etc.

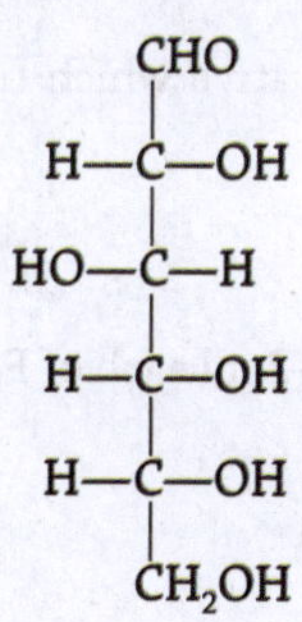

D(+)–Glucose

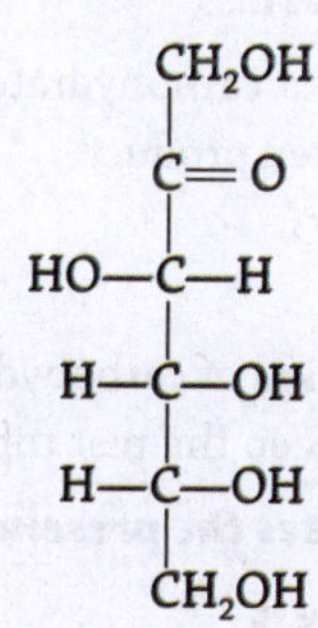

D(–)–Fructose

The carbohydrates which contain two to ten monosaccharide units are called **oligosaccharides.** For example, sucrose ($C_{12}H_{22}O_{11}$), maltose ($C_{12}H_{22}O_{11}$), raffinose ($C_{18}H_{32}O_{16}$), etc.

The carbohydrates which contain more than ten monosaccharide units are called **polysaccharides.** For example, starch, cellulose, glycogen, etc.

A more general classification of carbohydrates is into **sugars** and **non-sugars.** The **sugars** like glucose, fructose and canesugar are crystalline, water soluble and sweet substances. **Non-sugars** which include starch, cellulose, etc., are amorphous, insoluble in water and tasteless substances.

The carbohydrates which can reduce Tollen's reagent or Fehling solution are called **reducing sugars.** All monosaccharides are reducing sugars. Most of the disaccharides are also reducing sugars. **Sucrose is a non-reducing sugar.**

Carbohydrates are generally optically active because they contain chiral centres.

QUALITATIVE TESTS OF CARBOHYDRATES

EXPERIMENT 10.1

AIM: ***To study some simple tests of carbohydrates.***

REQUIREMENTS

Glucose, sucrose (cane-sugar), lactose (milk-sugar), starch, Molisch's reagent, Fehling's solution, Benedict's solution and iodine solution.

PROCEDURE

1. Molisch's Test

All carbohydrates give this test.

Take 1–2 ml of aqueous solution of carbohydrate (suspension in case of starch) and add few drops of Molisch's reagent (1% alcoholic solution of 1-naphthol). Put one ml of conc. H_2SO_4 slowly along the side of the test tube.

A red violet ring is produced at the junction of two layers.

Chemistry of the test

Conc. H_2SO_4 converts carbohydrates into furfural or its derivative which further reacts with 1-naphthol to give a coloured product.

2. Fehling's Test

Take 2 ml of aqueous solution of carbohydrate (nearly 5%) and add 1–2 ml each of Fehling's solution A and Fehling's solution B. Keep the test tube in boiling water bath.

Reddish ppt. indicates the presence of a reducing sugar.

Chemistry of the test

$$\underset{\text{Glucose}}{\begin{array}{c}CHO\\|\\(CHOH)_4\\|\\CH_2OH\end{array}} + 2Cu(OH)_2 + NaOH \xrightarrow[\text{ions}]{\text{Tartrate}} \underset{\substack{\text{Sod. salt of}\\ \text{gluconic acid}}}{\begin{array}{c}COO^-Na^+\\|\\(CHOH)_4\\|\\CH_2OH\end{array}} + 3H_2O + \underset{\text{red ppt.}}{Cu_2O\downarrow}$$

PREPARATION OF FEHLING'S SOLUTION

FEHLING'S SOLUTION A. Dissolve 17.5 g of $CuSO_4$ in 250 ml of distilled water containing few drops of H_2SO_4.

FEHLING'S SOLUTION B. Dissolve 86.5 g of sodium potassium tartarate and 30 g NaOH in 250 ml of distilled water.

3. Benedict's Test

To 1–2 ml of aqueous solution of carbohydrate in a test tube add 1–2 ml of Benedict's reagent. Keep the test tube in a boiling water bath.

Reddish ppt. indicates the presence of reducing sugar.

Note: *Chemistry of this test is the same as that of Fehling's test. Here citrate ions are used as complexing agent.*

PREPARATION OF BENEDICT'S REAGENT

Dissolve 17.3 g of sodium citrate and 10 g of anhydrous Na_2CO_3 in about 80 ml of distilled water. Heat if necessary. Dissolve 1.73 g of copper sulphate in 10 ml of water. Mix the two and make the volume 100 ml by adding water.

4. Tollen's Test

Take 2–3 ml of aqueous solution of carbohydrate in a test tube. Add to it 2–3 ml of Tollen's reagent. Keep the test tube in a boiling water bath for 10 minutes.

A shining silver mirror indicates the presence of reducing carbohydrate.

Chemistry of the test

$$AgNO_3 + NH_4OH \longrightarrow NH_4NO_3 + AgOH$$

$$2AgOH \longrightarrow Ag_2O + H_2O$$

$$Ag_2O + 2NH_4OH \longrightarrow \underset{\text{(Soluble)}}{2[Ag(NH_3)_2]OH} + 3H_2O$$

$$\underset{\text{Glucose}}{\begin{array}{c} CH_2OH \\ | \\ (CHOH)_4 \\ | \\ CHO \end{array}} + Ag_2O \xrightarrow{NH_4OH} \underset{\text{Gluconic acid}}{\begin{array}{c} COOH \\ | \\ (CHOH)_4 \\ | \\ COOH \end{array}} + \underset{\text{Silver mirror}}{2Ag}$$

PREPARATION OF TOLLEN'S REAGENT

Add NaOH solution to $AgNO_3$ solution. Then add NH_4OH solution dropwise till the ppt. just dissolve. The clear solution obtained is Tollen's reagent.

5. Iodine Test (For starch only)

To the aqueous suspension of the sample, add 1–2 drops of iodine solution.

Appearance of blue colouration indicates the presence of starch.

OBSERVATIONS

Test	*Glucose*	*Lactose*	*Sucrose*	*Starch*
1. Taste	Sweet	Sweet	Sweet	Tasteless
2. Solubility	Soluble	Soluble	Soluble	Insoluble
3. Molisch's test	Purple ring	Purple ring	Purple ring	Purple ring
4. Fehling's test	Red ppt.	Red ppt.	Negative	Negative
5. Benedict's test	Red ppt.	Red ppt.	Negative	Negative
6. Iodine test	Negative	Negative	Negative	Blue colour

10.02. OILS AND FATS

Chemically fats and oils are triesters of glycerol and higher fatty acids. At ordinary temperature oils are liquids while fats are solids. As compared to fats oils contain a large proportion of unsaturated acid radicals. Fats and oils are of vegetable or animal origin. These serve as excellant *source of energy for the body* as by combustion they produce heat and energy. They *form fatty tissues around delicate organs* to protect them from injury. They also *form a heat insulating coat around the body.*

QUALITATIVE TESTS FOR OILS AND FATS

EXPERIMENT 10.2

AIM: ***To study some simple tests of oils and fats.***

REQUIREMENTS

Chloroform, ethyl alcohol, $KHSO_4$ crystals, furfural solution, Huble's solution, *desi ghee*, vegetable *ghee* and refined oil.

PROCEDURE

1. Solubility Test

This test is based on the fact that oils and fats are soluble in organic solvents like chloroform, alcohol, etc., but are insoluble in water. Shake a small amount of given sample with 5 ml each of water, alcohol and chloroform in three test tubes and observe the solubility and draw inferences as follows:

Test tube	*Solvent*	*Observations*	*Inference*
1.	Water	(*i*) Sample is *immiscible.*	Oil or fat present.
		(*ii*) Sample is *miscible.*	Oil or fat absent.
2.	Alcohol	(*i*) Sample forms lower layer, which *dissolves on heating.*	Oil or fat present.
		(*ii*) Sample *does not dissolve* even on heating.	Oil or fat absent.
3.	Chloroform	(*i*) Sample is *miscible.*	Oil or fat present.
		(*ii*) Sample is *immiscible.*	Oil or fat absent.

2. Transluscent Spot Test

Press a little of the substance in the folds of the filter paper. On unfolding the filter paper, the appearance of **transluscent or greasy spot** on the filter paper indicates the presence of fat or oil. The spot grows larger on heating and drying the filter paper.

3. Acrolein Test

Heat a little of the sample with some crystals of $KHSO_4$ in a test tube.

A pungent irritating odour of acrolein confirms the presence of fat or oil.

$$\text{Oil or fat} \xrightarrow{\Delta} \text{Glycerol} + \text{Fatty acid}$$

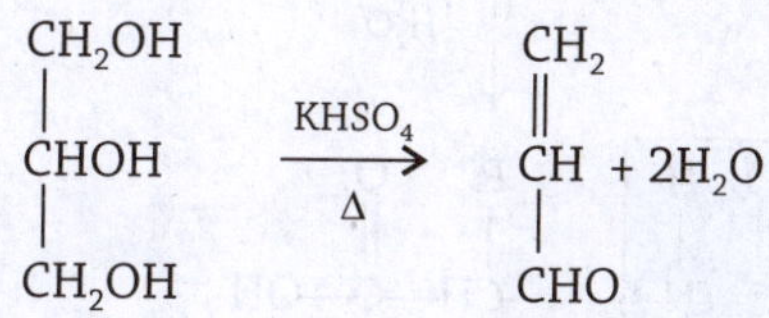

Glycerol   Acrolein

4. Baudouin Test

This test is applied to distinguish between desi ghee and vanaspati ghee.

Vanaspati ghee contains 5% seasame oil. Pure desi ghee does not contain seasame oil.

Shake 5 ml of melted ghee with 5 ml of conc. HCl and 2–3 drops of 2% furfural solution in alcohol. Keep it aside for 5–10 minutes. **Rose red colour appears if seasame oil is present.**

This test can be applied to find out whether the given sample of desi ghee contains vanaspati ghee or not.

5. Huble's Test

This test is applied to know degree of unsaturation in the given sample of oil or fat.

Take two test tubes, label them as I and II. Put in each test tube 3 ml of chloroform. Add 3–4 drops of cotton seed oil in test tube I and linseed oil in test tube II. Shake and add 3 drops of Huble's reagent in each test tube and observe the fading of violet colour in test tubes. The violet colour of iodine fades away in test tube II, while, violet colour in test tube I does not fade away. This indicates that linseed oil is more unsaturated than cotton seed oil.

PREPARATION OF HUBLE'S SOLUTION

Mix equal volumes of 5 to 7% $HgCl_2$ in alcohol with 5% solution of iodine in 96% alcohol.

10.03. PROTEINS

Proteins are high molecular mass, long chain polymers composed of α-amino acids.

Amino acids are molecules that have both an amino (NH_2) and a carboxylic (COOH) group. The **amino acids** in proteins are called α-amino acids because they have the amino group attached to the α-carbon atom. α-Amino acids exist as zwitter ions and are crystalline solids.

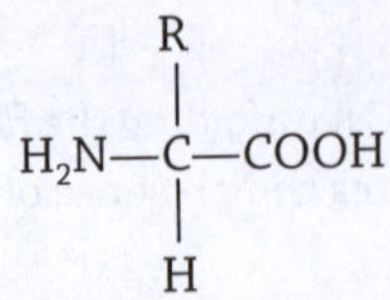

α-Amino acids

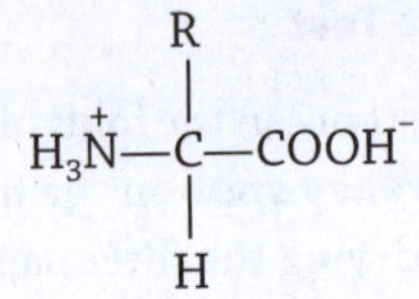

Zwitter ion

The amino acids contain an acidic group and a basic group. They undergo condensation as shown below:

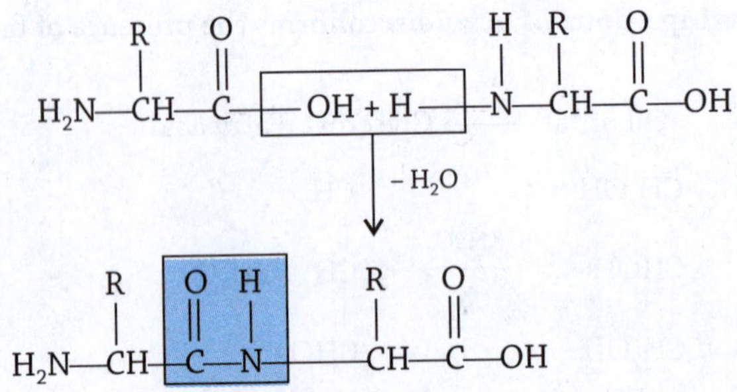

Peptide linkage

The $-\overset{O}{\overset{\|}{C}}-NH-$ linkage that joins the two amino acid units is called **peptide linkage.** The product formed by the combination of two α-amino acid molecules is called **dipeptide** and with three α-amino acid molecules is called **tripeptide**. A polypeptide contains large number of α-amino acid molecules. *A polypeptide having molecular mass greater than 10000 is called protein.*

QUALITATIVE TESTS OF PROTEINS

EXPERIMENT 10.3

AIM: ***To study some simple tests of proteins.***

REQUIREMENTS

Egg albumin dispersion, gelatin dispersion, Millon's reagent and Ninhydrin reagent.

PROCEDURE

1. Biuret Test

To the dispersion of the substance to be tested (say 5% solution of egg albumin) add about 2 ml of NaOH solution. Now add 4–5 drops of 1% $CuSO_4$ solution. Warm the mixture for about five minutes.

Bluish violet colouration indicates the presence of protein.

2. Xanthoproteic Test

Take about 2 ml of egg albumin dispersion in a test tube and add a few drops of conc. HNO_3 and heat.

A yellow colouration indicates the presence of proteins.

3. Ninhydrin Test

Take about 2 ml of egg albumin dispersion in a test tube and add 3–4 drops of Ninhydrin solution. Boil the contents.

Intense blue colouration confirms the presence of proteins.

NINHYDRIN SOLUTION is prepared by dissolving 0.1 g of ninhydrin in about 100 ml of distilled water. This solution is unstable and can be kept only for two days.

4. Millon's Test

This test is given by proteins containing phenolic amino acids. *Gelatin does not give this test.*

To 1–2 ml of egg albumin dispersion add 2 drops of Millon's reagent.

White ppt. which changes to brick red on boiling, confirms the presence of proteins.

MILLON'S REAGENT is prepared by dissolving 5 g each of $HgNO_3$ and $Hg(NO_3)_2$ in 100 ml of dil. HNO_3.

TEST FOR CARBOHYDRATES, FATS AND PROTEINS IN FOOD STUFFS

EXPERIMENT 10.4

AIM: ***To detect the presence of carbohydrates, fats and proteins in the following food stuffs: Grapes, potatoes, rice, butter, biscuits, milk, groundnut, boiled egg.***

THEORY

The presence of carbohydrates, fats and protein in any food stuff is detected by performing the tests for carbohydrates, fats and proteins with the extract of the food stuff. These tests do not interfere with each other.

APPARATUS

Test tubes, beakers, glass rod, pestle mortar and burner.

PROCEDURE

First prepare the extract of the given food stuff by either dry grinding in the mortar with a pestle or by boiling with minimum quantity of water extracting with a small quantity of an organic solvent after grinding the food stuff. The extracts of some of the food stuffs can be prepared as given under:

For grapes — extract the juice
For potatotes — cut into slices and boil with water
For rice — boil with water
For butter — test directly
For biscuits — boil with water
For milk — test directly

For groundnut — grind in the mortar

For boiled egg — take the white of the egg, grind and shake with water.

With the solution or the suspension perform tests for carbohydrates, fats and proteins as discussed in Expts. 10.1, 10.2 and 10.3.

OBSERVATIONS

Food Stuff	For Carbohydrates				For Proteins		For Fats and Oils		
	Molisch's Test	Fehling Test	Tollen's Test	Iodine Test	Biuret Test	Ninhydrin Test	Solubility Test	Spot Test	Acrolein Test
1. Grapes									
2. Potatoes									
3. Rice									
4. Butter									
5. Biscuits									
6. Milk									
7. Groundnut									
8. Boiled egg									

RESULT

The given food stuff contains

(*i*) .... (*ii*) .... (*iii*) ....

PRECAUTIONS

1. Use freshly prepared reagents for performing tests.
2. Use minimum quantities of reagents.

VIVA QUESTIONS WITH ANSWERS

Q. 1. What are carbohydrates?

Ans. Carbohydrates are polyhydroxy aldehydes or polydroxy ketones or the compounds which yield these on hydrolysis.

Q. 2. Give two examples of monosaccharides.

Ans. Glucose and fructose.

Q. 3. Give two examples of disaccharides.

Ans. Sucrose and maltose.

Q. 4. What is Tollen's reagent?

Ans. A solution prepared by adding NaOH solution to $AgNO_3$ solution and then adding NH_4OH to dissolve the ppt.

Q. 5. What are the functions of carbohydrates?

Ans. 1. To supply energy to the body and to act as storage of chemical energy in the form of glycogen in liver.

2. As a constituent of cell membrane.

Q. 6. Why do we get a red ppt. in Fehling's test?

Ans. Because of formation of cuprous oxide (Cu_2O).

Q. 7. Why do we get a shining mirror in Tollen's test?

Ans. Due to formation of silver by reduction of Ag^+ ions by aldehydes, which deposits on the walls of the test tube.

Q. 8. What is Molisch's reagent?

Ans. Alcoholic solution of α-naphthol.

Q. 9. What is purple ring formed due to, in Molisch's test?

Ans. Conc. H_2SO_4 converts carbohydrates into furfural or its derivative which then reacts with α-naphthol to form a violet coloured compound.

Q. 10. Name some reducing and non-reducing sugars.

Ans. Glucose and fructose are examples of reducing sugar and sucrose is an example of non-reducing sugar.

Q. 11. How will your distinguish between sucrose and glucose?

Ans. Glucose, being a reducing sugar, will give silver mirror test positive (Sucrose is a non reducing sugar).

Q. 12. What is the role of tartarate ions in Fehling's reagent?

Ans. It acts as complexing agent and prevents the precipitation of copper (II) hydroxide.

Q. 13. What is the role of citrate ions in Benedict's solution?

Ans. It acts as complexing agent and prevents the precipitation of copper (II) hydroxide.

Q. 14. Why does fructose reduce Fehling's solution and Tollen's reagent though it contains ketonic group and not aldehydic group?

Ans. In alkaline medium fructose rearranges to glucose and the two are equilibrium with each other (Lobry de Bruyn-van Ekenstein rearrangement).

Q. 15. What are Fehling A and Fehling B solutions?

Ans. Fehling solution A is copper sulphate solution and Fehling solution B is mixture of solutions of sodium potassium tartrate and sodium hydroxide.

Q. 16. Do all the sugars give Fehling solution test?

Ans. No, only reducing sugars like glucose and fructose give this test.

Q. 17. What are proteins?

Ans. Proteins are naturally occurring complex nitrogenous organic substances with high molecular masses. Chemically, they are polypeptides formed by the condensation of α-amino acids.

Q. 18. What is the colour obtained in Ninhydrin test for proteins?

Ans. Blue.

Q. 19. How are proteins affected by heat?

Ans. They undergo coagulation.

Q. 20. How are proteins affected by conc. HNO_3?

Ans. Turn yellow.

Q. 21. What is the name given to the reaction between protein and conc. HNO_3?

Ans. Xanthoprotein reaction.

Q. 22. What are the final products of hydrolysis of proteins?

Ans. α-Amino acids.

Q. 23. What is biuret test for proteins?

Ans. To 2–3 ml of protein solution in a test-tube, add an equal volume of 10% NaOH solution. Mix thoroughly and add a few drops of 0.5% copper sulphate solution. A **purple-violet colour** is obtained, if protein is present.

Q. 24. What are oils and fats?

Ans. Oils and fats are triesters of glycerol with higher fatty acids. They are also called triglycerides.

Q. 25. What is the difference between oils and fats?

Ans. Oils are liquids at room temperature. They contain higher proportion of unsaturated acids whereas fats are solids at room temperature and contain higher proportion of saturated acids.

Q. 26. Name two tests for testing fats or lipids.

Ans. (*i*) Solubility test (*ii*) Acrolein test.

CHAPTER 11

Volumetric Analysis

In volumetric analysis, the quantities of the constituents present in the given unknown solution are determined by measuring the volumes of the solutions taking part in the given chemical reaction. The main process of this analysis is called titration which means the determination of the volume of a reagent required to bring a definite reaction to completion.

11.01. APPARATUS USED IN VOLUMETRIC ANALYSIS

In volumetric analysis, the apparatus required is as follows:

(i) Graduated burette, pipette, measuring flasks and measuring cylinders.

(ii) Titration flasks, beaker, tile, glass rod, funnel, weighing bottle, wash bottle.

(iii) A chemical balance for weighing.

USING A BURETTE

It is a long, cylindrical tube of uniform bore fused at the lower end with a stop-cock (Fig. 11.1). It is graduated in millilitres from 0 to 50. Each division is further sub-divided into ten equal parts. Therefore, each sub-division reads 0.1 ml.

Before a burette is filled with the solution, it is thoroughly washed, so that no greasy matter is sticking inside or outside the burette. **No drops should adhere to the inner wall of a clean burette.** Take a small volume of solution to be taken in it, close the upper mouth of the burette with the thumb and hold in horizontal position as shown in Fig. 11.2. Rotate the burette so as to wet the inner walls of the burette. Reject this solution through the stop-cock. *This process is known as* **rinsing.** Then the burette is filled with the help of a funnel inserted in the top Fig. 11.3. The funnel must then be taken out after filling the burette. **The solution in the burette is called titrant.**

Care must be taken that no air bubbles remain in the narrow bottom tip of the burette. To remove this air, the stop-cock is opened and the liquid is allowed to run out rapidly into the beaker or flask.

Burette reading forms the most important aspect of the experiment, therefore, burette should be read very carefully, after removing parallax.

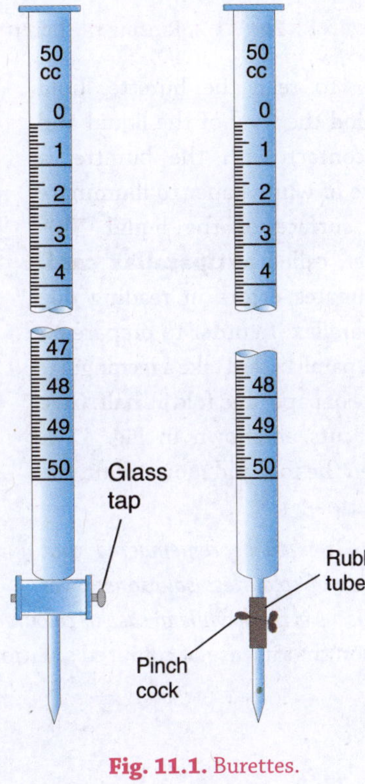

Fig. 11.1. Burettes.

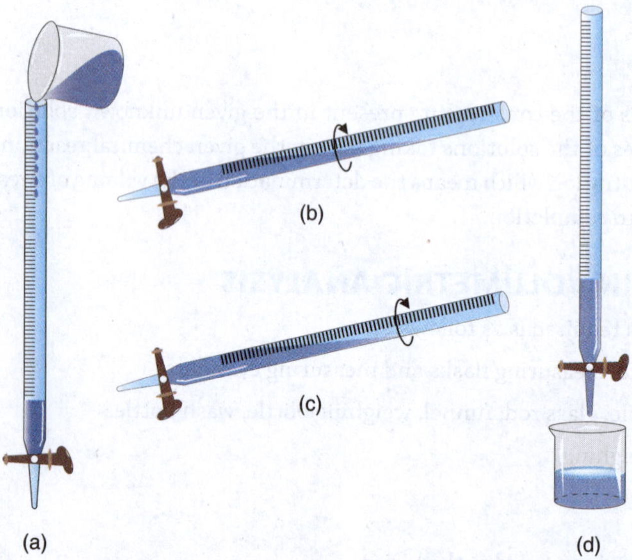

Fig. 11.2. Rinsing the burette.

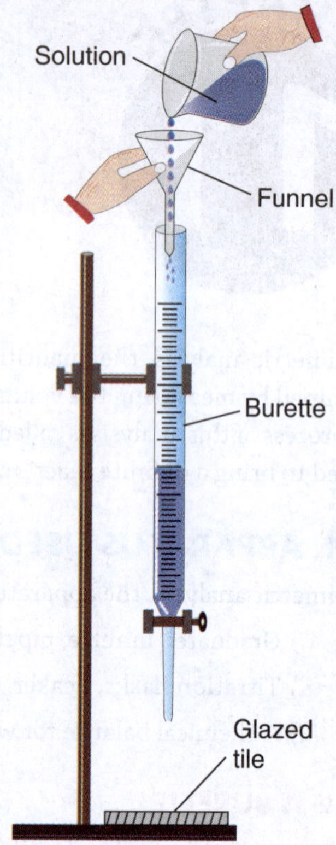

Fig. 11.3. Filling the burette.

To read the burette, hold behind the level of the liquid and in contact with the burette, a piece of white paper to illuminate the surface of the liquid. This paper, called **antiparallax card,** eliminates errors in reading due to parallax. In order to prepare an anti-parallax card take a rectangular piece of paper and fold it half. Give two cuts as shown in Fig. 11.4. Open the fold and mount it on the burette.

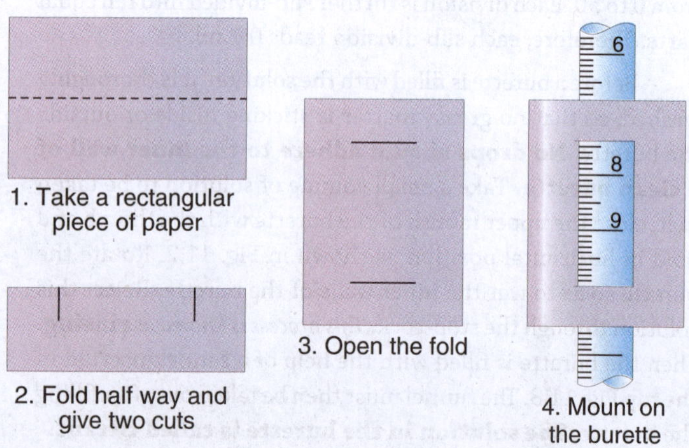

Fig. 11.4. Making and mounting an antiparallax card.

It is to be remembered that in case of colourless solutions lower meniscus is read, while in case of coloured solutions, level is read from the upper meniscus. This is due to the reason that in case of coloured solutions lower meniscus is not visible clearly.

Take reading of the burette placing your eye exactly in front of meniscus (Fig. 11.6) of the solution.

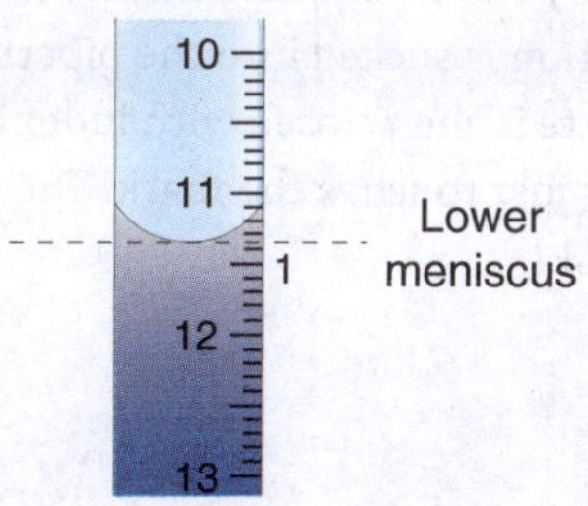

Fig. 11.5. Lower meniscus.

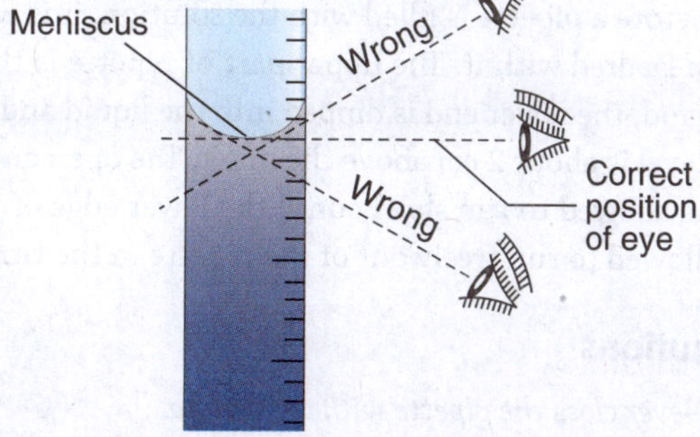

Fig. 11.6. Correct way of reading burette.

Precautions

1. *See that stop-cock does not leak.*
2. *Remove the funnel immediately after filling the burette.*
3. *Do not allow any air bubble to remain inside the burette.*
4. *Always use antiparallax card and place the eye exactly in the level of meniscus.*
5. *Let no drops of solution be hanging at the tip of the burette at the end point.*

USING A PIPETTE

This apparatus is used for accurate measurements of definite volume of solution. It consists of a long narrow tube with cylindrical bulb in the middle and a jet at its lower end.

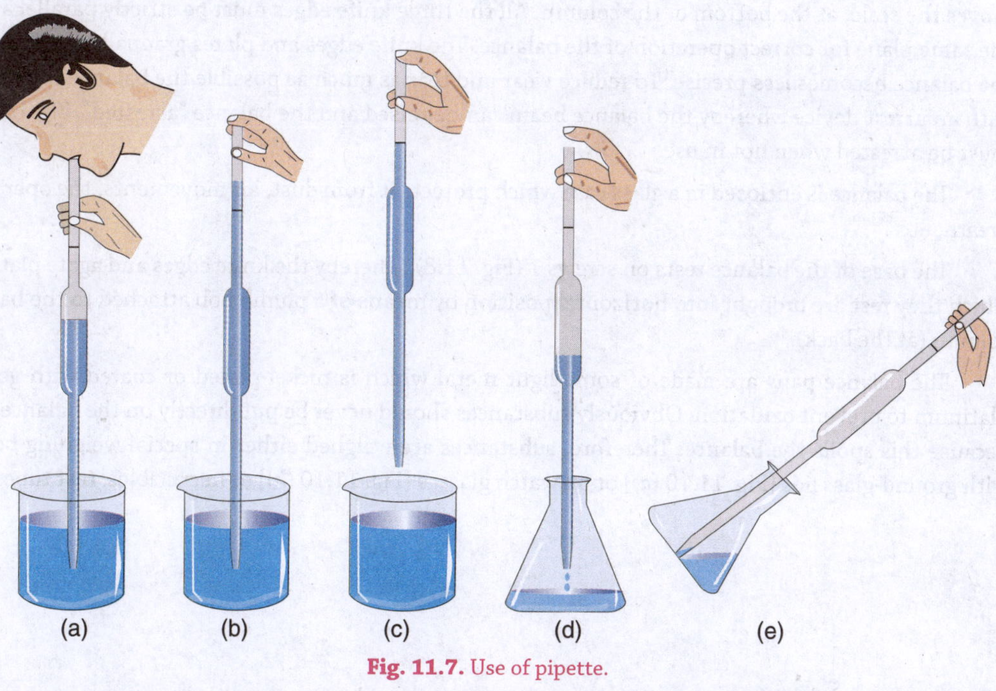

Fig. 11.7. Use of pipette.

On the upper part of the stem, there is an etched circular mark. On the bulb is marked the volume which the pipette can deliver when filled up to the circular mark [Fig. 11.7 (*a*)].

Before a pipette is filled with the solution, it is washed and thoroughly rinsed with the solution to be measured with it. The upper part of pipette is then held by the thumb and middle finger of the right hand, the lower end is dipped into the liquid and the solution is sucked into the pipette until the liquid level is about 2 cm above the mark. The open end of pipette is then closed with index finger. The liquid is allowed to run slowly until the lower edge of meniscus just touches the mark. The solution is then allowed to run freely out of the pipette in the titration flask.

Precautions

1. *Never close the pipette with the thumb.*
2. *Keep the lower end always dipping in the liquid while sucking the liquid.*
3. *Never pipette out hot solutions or corrosive solutions.*
4. *Do not blow out the last drop of the solution from the jet end.*

11.02. CHEMICAL BALANCE

The balance is the principal instrument used in quantitative analysis. One of the most important requirements in quantitative analysis is a sufficiently high degree of precision. The analytical balance used in quantitative analysis can be used for weighing objects not heavier than 100–200 g to a precision of 0.0002 g, *i.e.*, 0.2 mg. The most usual design of a balance of this type is shown in Fig. 11.8.

The most important part, the beam, has three knife edges made of agate or very hard steel [Fig. 11.9 (*a*)]. The central knife edge rests on a special very smooth agate plate on the top of the balance column. The balance pans are suspended from the terminal knife edges by means of stirrups [Fig. 11.9 (*b*)].

A pointer is fixed to the centre of the beam; as the balance swings the lower end of the pointer moves the scale, at the bottom of the column. All the three knife edges must be strictly parallel and in the same plane for correct operation of the balance. The knife edges and plates gradually wear out and the balance becomes less precise. To reduce wear and tear as much as possible the balance is provided with an arrest device whereby the balance beam can be raised and the balance "arrested". The balance must be arrested when not in use.

The balance is enclosed in a glass case which protects it from dust, air movements, the operator's breath, etc.

The base of the balance rests on screws 1 (Fig. 11.8), whereby the knife edges and agate plates on which they rest are brought into horizontal position by means of a plumb bob attached to the balance column (at the back).

The balance pans are made of some light metal which is nickel-plated or coated with gold or platinum to prevent oxidation. Obviously substances should never be put directly on the balance pans because this spoils the balance. Therefore, substances are weighed either in special weighing bottles with ground-glass lids [Fig. 11.10 (*a*)] or on watch glasses [Fig. 11.10 (*b*)] or in crucibles, test tubes, etc.

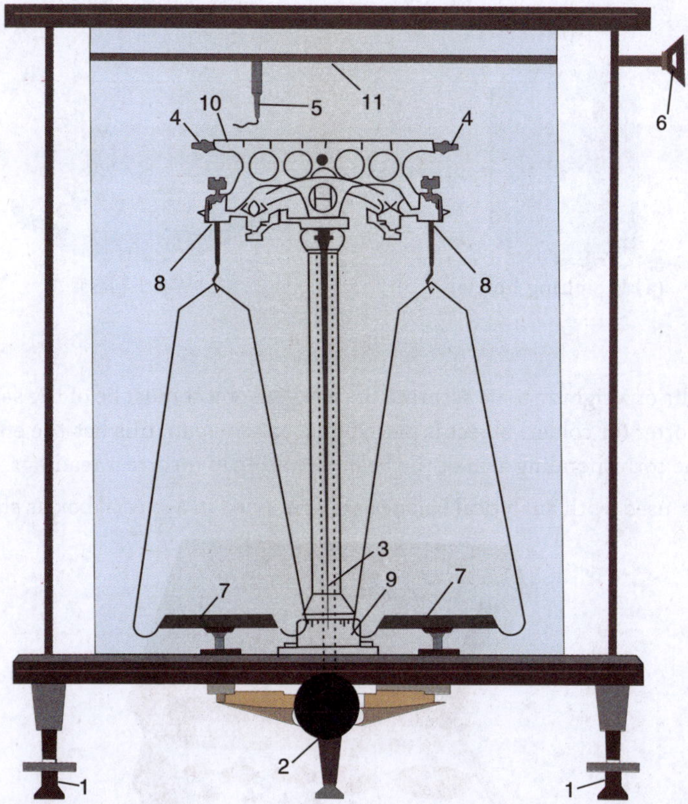

1. Adjusting screws; 2. Arrest knob; 3. Pointer;
4. Screws for zero point adjustment; 5. Rider hook;
6. Knob of rider carrier; 7. Balance pans; 8. Stirrups;
9. Scale; 10. Graduated beam; 11. Rider carrier.

Fig. 11.8. Analytical balance.

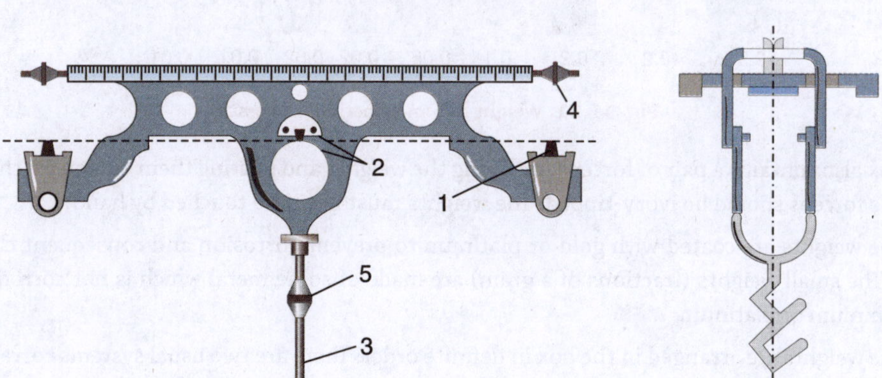

1. Terminal knife edge; 2. Central knife edge; 3. Pointer;
4. Screw for zero point adjustment; 5. Weight for sensitivity adjustment.

Fig. 11.9. Parts of analytical balance.

(a) Weighing bottle. (b) Watch glass.

Fig. 11.10

For the results of weighing to be accurate the weighed object must be of the same temperature as the balance. If a hotter (or colder) object is placed on a balance pan, this has the effect of lengthening (or shortening) the corresponding arm of the beam resulting in incorrect readings.

The weights used with analytical balance are contained in a special box as shown in Fig. 11.11.

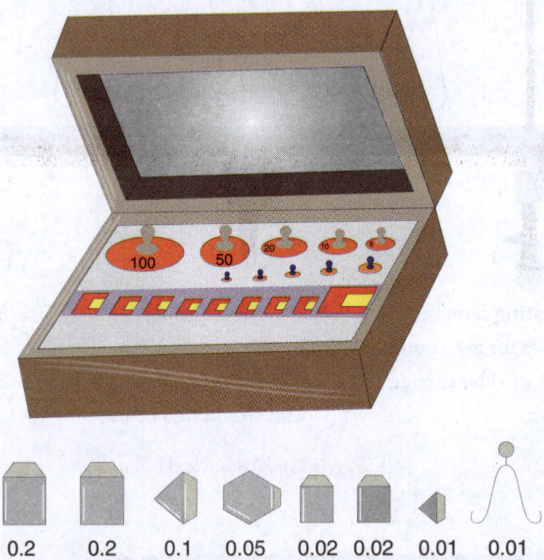

Fig. 11.11. Weight box and fractional weights.

Box also contains a pair of forceps for lifting the weights and putting them on and off the balance pans. The forceps should be ivory-tipped. The weights must never be touched by hand.

The weights are coated with gold or platinum to prevent corrosion and consequent changes of weight. The small weights (fractions of a gram) are made of some metal which is not corroded in air, *e.g.*, aluminium or platinum.

The weights are arranged in the box in definite order. There are two usual systems corresponding to the numbers 5 : 2 : 2 : 1 or 5 : 2 : 1 : 1. In accordance with the first system, the box would contain weights of 50, 20, 20, 10, 5, 2, 2, 1 g and in accordance with the second, weights of 50, 20, 10, 10, 10, 5, 2, 1, 1, 1 g. Fractions of a gram follow the same systems and are made of different shapes so that small weights are easier to distinguish. For example, fractional weights of 0.5 and 0.05 g are made in shape of regular hexagon, weights 0.2 and 0.02 g are squares and weights 0.1 and 0.01 g are triangles. Each fractional weight has an edge bent at right angle by which it is lifted with the forceps.

By means of the weights an object can be weighed to an accuracy of 0.01 g. Thousandth and ten-thousandth fractions of a gram are weighed by means of the so called rider. The rider, as shown in Fig. 11.12, is a thin bent wire (usually of aluminium) weighing 0.01 g or 0.005 g, it is attached with the aid of the forceps by its loop on hooks. This hook is fixed to the horizontal rod 11 with the knob 6 outside the balance case. This rod is rotated or moved to place the rider at any desired point on the beam. The beam has a scale, the graduations of which differ in different balances. If the rider is moved from the zero division to the fifth (*i.e.*, exactly over the central knife edge), this is equivalent to removal of 0.005 g from the left-hand pan or a similar increase of the load on the right-hand pan.

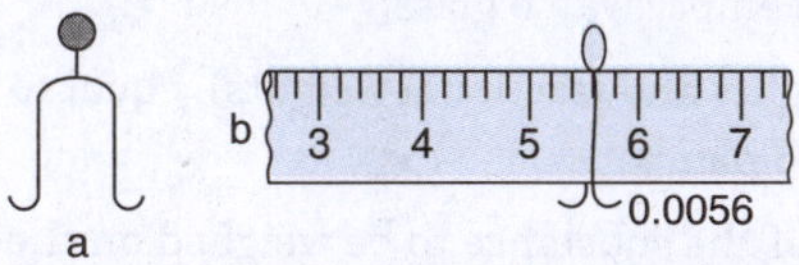

Fig. 11.12. Rider and readings along the balance beam.

Setting the Balance

Before the substance can be weighed in a balance it has to be first set in proper order. The following steps are followed for setting the balance:

1. Clean the pans of the balance with a hair-brush or a clean handkerchief.
2. Level the balance by adjusting the levelling screws. See that the pointer rests at zero. Close the front door of the balance.

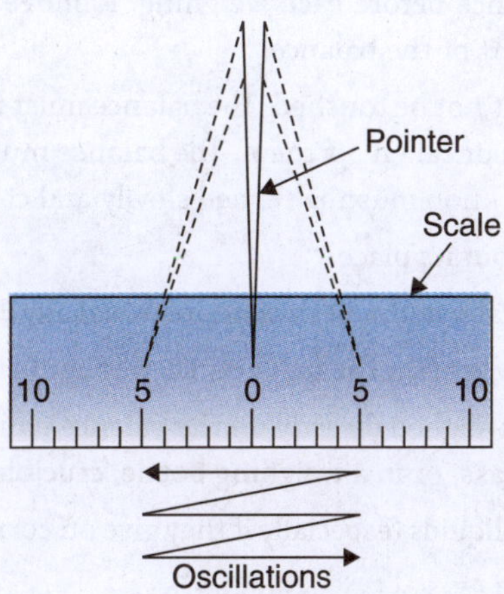

Fig. 11.13

3. Now rotate the key arrest knob to raise the beam and see that the pointer swings or oscillates equal divisions on both sides of the zero mark as shown in Fig. 11.13. If it does not oscillate equally on both the sides arrest the beam and move the adjusting screws (4) till on rotating the arrest knob, the pointer oscillates equally on both sides of the zero mark. Again arrest the beam.

Weighing the Substance

1. Take a clean and dry watch glass or weighing bottle and place it carefully on the left hand pan of the balance.
2. Pick out an appropriate gram weight from the weight box with the help of forceps and place it on the right hand pan. If the gram weight is heavier as compared to the weight of the watch glass, remove it and try lower weight. The gram weight should be slightly less than the weight of the watch glass (less than 1 gram).
3. After placing the correct gram weight start placing fractional weights.
4. Use rider for weights lighter than 10 mg.
5. Record the correct weight of empty watch glass.
6. Now add weights (gram weights and fractional weights), equal to the amount of the substance to be taken, in the right hand pan.
7. Now add required quantity of the substance to be weighed on the watch glass.
8. Take out the watch glass along with the substance.
9. Clean the balance and close it.

Precautions While Handling the Analytical Balance

In weighing it must be remembered that the analytical balance is a precise physical instrument which must be handled with great care.

To avoid damage to the balance and to ensure accurate weighing the following rules must be strictly observed:

1. Check the state of the balance before each weighing. Remove dust from the pans with a soft brush and find the zero point of the balance.
2. The unarrested balance must not be touched. The balance must be arrested before the object and weights are put on the pans or taken off them. The balance must be arrested before the rider is moved along the beam. The knob must be turned slowly and carefully.
3. Do not move the balance from its place.
4. Never overload the balance above the permitted load (usually 100 g) as this causes damage.
5. Do not place wet or dirty objects on the balance. Do not spill anything inside the balance case.
6. Do not put the object to be weighed directly on the balance pan. Do not use pieces of paper; put the substance on a watch glass, or in a weighing bottle, crucible, test tube, etc.
7. Hygroscopic substances and liquids (especially if they give off corrossive vapours) must be weighed in closed weighing bottles.
8. Do not weigh hot (or very cold) objects. The object to be weighed must reach the temperature of the balance. It must, therefore, be left for at least 20 minutes in a dessicator near the balance.
9. Always use only the side doors of the balance case when weighing. The front door, must be kept shut all the time.
10. Do not touch the balance, weights or rider with the fingers. The weights must be handled by special forceps.

11. Do not muddle the weights. Each weight must be put in its proper place in the box.
12. Remain in the balance room only while weighing.

11.03. ELECTRONIC BALANCE

The modern instrument for weighing an object in the laboratory is the electronic balance. The electronic balance provides convenience in weighing and is much more accurate than mechanical balance. With an electronic balance, a single on-off operation permits the operator to read the weight of an object on the balance pan immediately from a digital display. The majority of electronic balances incorporate **tare** facility which permits the weight of a container to be cancelled out, so that when material is added to the container, the weight recorded is simply the weight of the material. Most balances of this type have a self-testing system which indicates the balance is functioning correctly each time it is switched on.

Fig. 11.14. Electronic balance

The balance pan in an electronic balance is situated within an enclosure provided with glass doors which can be opened to provide access to the pan.

Depending upon the weight range and accuracy different types of electronic balances are available.

Macrobalances can weigh upto about 200 g and reading to 0.1 mg.

Semimicrobalances can weigh upto about 30 g and reading to 0.01 mg.

Microbalances can weigh upto about 20 g and reading to 1 μg.

Ultramicrobalances can weigh upto about 5 g and reading to 0.1 μg.

The weighing process with an electronic balance includes the following steps.

1. Brush the balance pan lightly to remove any dust.
2. Keeping the balance at rest, place the object on the pan and close the pan compartment.
3. Set the on-off control of the balance to the 'on' position. Observe the value shown on the digital display and record it in the notebook.
4. Return the on-off control of the balance to the 'off' position.
5. After completing weighing, remove the object which has been weighed.
6. Clear up any accidental spillages and close the pan compartment.

Precautions while using Electronic Balances

1. Keep the balance clean. Remove dust from the pan and the floor of the pan compartment with a camel-hair brush.
2. Objects to be weighed should never be handled with the fingers; use tongs or a loop of clean paper.
3. Objects to be weighed should be allowed to attain the room temperature before weighing.
4. Do not exceed the stated maximum load of the balance.
5. Chemicals should not be directly placed on the balance pan.
6. Avoid exposing the balance to corrosive atmospheres.

11.04. SOME IMPORTANT TERMS

1. Standard Solution

A solution whose concentration is known, is called a standard solution. Concentration of a solution is generally expressed in terms of normality or molarity.

2. Normality

Normality of a solution is defined as the number of gram-equivalents of solute per litre of solution. It is denoted by N. Mathematically, it may be expressed as:

$$\text{Normality} = \frac{\text{Number of gram-equivalents of solute}}{\text{Volume of solution (in litres)}}$$

or $$= \frac{\text{Mass of solute (in grams) per litre of solution}}{\text{Gram-equivalent mass of the solute}}$$

$\therefore$ Number of gram equivalents of solute = Normality × Volume of solution (in litres).

A solution containing one gram-equivalent of solute per litre of solution is called **normal solution.**

3. Molarity

Molarity of a solution may be defined as the number of gram moles of solute per litre of the solution. It is denoted by M. Mathematically, it may be expressed as:

$$\text{Molarity} = \frac{\text{Gram moles of solute}}{\text{Volume of solution (in litres)}}$$

or $$= \frac{\text{Mass of solute (in grams) per litre of solution}}{\text{Gram molecular mass of the solute}}$$

$\therefore$ Gram moles of solute = Molarity × Volume of solution (in litres)

A solution containing one gram mole of solute per litre of solution is called **molar solution.**

4. End Point

It is the point where the reaction between the two solutions is just complete.

5. Indicator

A substance which indicates the attainment of end point. Indicator undergoes a change in colour at the end point.

11.05. EQUIVALENT MASSES OF OXIDIZING AND REDUCING AGENTS

According to electronic concept, oxidation *is the process which results in the loss of one or more electrons by atoms or ions* and **reduction** *is the process which results in the gain of one or more electrons by atoms or ions.* The **oxidising agent** *is the substance which gains one or more electrons and gets reduced.* The **reducing agent** *is the substance which loses one or more electrons and gets oxidised.*

The equivalent mass of an oxidising agent is equal to its molecular mass (or formula mass) divided by the number of the electrons gained by one molecule or ion of the substance in the reaction.

$$\therefore \quad \text{Equivalent mass of an oxidising agent} = \frac{\text{Molecular mass or formula mass}}{\text{No. of electrons gained by one molecule}}$$

$$\text{Equivalent mass of a reducing agent} = \frac{\text{Molecular mass or formula mass}}{\text{No. of electrons lost by one molecule}}$$

Equivalent Mass of Potassium Permanganate, $KMnO_4$

In the acidic medium, permanganate (active ion) is reduced as follows:

$$MnO_4^- + 8H^+ + 5e^- \longrightarrow Mn^{2+} + 4H_2O.$$

Here, the number of electrons gained by one permanganate is 5.

$$\therefore \quad \text{Equivalent mass of } KMnO_4 \text{ in acidic medium} = \frac{\text{Molecular mass}}{5} = \frac{158}{5} = 31.6.$$

Equivalent Mass of Ferrous Ammonium Sulphate, $FeSO_4(NH_4)_2SO_4.6H_2O$ (Mohr's salt)

$$FeSO_4(NH_4)_2SO_4.6H_2O \equiv FeSO_4 + (NH_4)_2SO_4 + 6H_2O$$

$$FeSO_4 \rightleftharpoons Fe^{2+} + SO_4^{2-}$$

Fe^{2+} is oxidised to Fe^{3+} by losing one electron

$$Fe^{2+} \longrightarrow Fe^{3+} + e^-$$

Number of electrons lost per molecule of Mohr's salt is 1.

$\therefore$ Equivalent mass of Mohr's salt $FeSO_4.(NH_4)_2SO_4.6H_2O$

$$= \frac{\text{Molecular mass}}{1} = \frac{392}{1} = 392.$$

Equivalent Mass of Oxalic Acid

During reaction with acidified $KMnO_4$, oxalic acid is oxidised to CO_2.

Here the reaction takes place as:

$$\begin{array}{l}COOH\\ |\\ COOH\end{array} \longrightarrow 2CO_2 + 2H^+ + 2e^-$$

The number of electrons lost per molecule of oxalic acid = 2.

$$\therefore \quad \text{Eq. mass of crystalline oxalic acid, } \begin{array}{l}COOH\\ |\\ COOH\end{array} . 2H_2O$$

$$= \frac{\text{Molecular mass of crystalline oxalic acid}}{2}$$

$$= \frac{126}{2} = 63$$

The equivalent masses of some common substances which we come across during redox titrations at this level are given in Table 11.1.

Table 11.1. Molecular Masses and Equivalent Masses of Some Substances

Substance	Molecular mass	Ionic equation	n	Eq. mass = $\frac{\text{Mol. mass}}{n}$
Pot. permanganate ($KMnO_4$)	158	$MnO_4^- + 8H^+ + 5e^- \longrightarrow Mn^{2+} + 4H_2O$	5	31.6
Mohr's salt [$FeSO_4(NH_4)_2SO_4.6H_2O$]	392	$Fe^{2+} \longrightarrow Fe^{3+} + e^-$	1	392
Ferrous sulphate (anhydrous) ($FeSO_4$)	152	$Fe^{2+} \longrightarrow Fe^{3+} + e^-$	1	152
Ferrous sulphate (crystals) ($FeSO_4.7H_2O$)	278	$Fe^{2+} \longrightarrow Fe^{3+} + e^-$	1	278
Oxalic acid (anhydrous) ($H_2C_2O_4$)	90	$C_2O_4^{2-} \longrightarrow 2CO_2 + 2e^-$	2	45
Oxalic acid (crystals) ($H_2C_2O_4.2H_2O$)	126	$C_2O_4^{2-} \longrightarrow 2CO_2 + 2e^-$	2	63

11.06. PREPARING A STANDARD SOLUTION

A standard solution is prepared by dissolving a definite weight of substance (a primary standard), in a definite volume. A substance is classified as a **primary standard** if it has following characteristics:

1. *It is easily available in state of high purity.*
2. *It is neither hygroscopic nor deliquescent.*
3. *It shows high solubility in water.*
4. *It does not dissociate or decompose during storage.*
5. *It should react instantaneously with another substance in stoichiometric proportion.*

Substances whose standard solutions cannot be prepared directly are called **secondary standard substances.** These include those substances which are not available in the pure form. For example, potassium permanganate, NaOH, KOH, etc. The solutions of secondary standards are standardized by titrating against solution of some primary standard.

For preparing a standard solution, student must remember that he is working on precise experiments, where the slightest inaccuracy may distort the analytical results which may have taken a great deal of work and time to obtain. It is, therefore, specially important to follow strictly the usual rules concerning orderly and clean work. The apparatus required for a given determination must be procured before hand and washed thoroughly and the weighing must be done accurately.

EXPERIMENT 11.1

AIM: ***Prepare 250 ml of 0.1 M solution of oxalic acid from crystalline oxalic acid.***

THEORY

Molecular mass of crystalline oxalic acid $\left(\begin{array}{l}\text{COOH}\\ |\\ \text{COOH}\end{array} . 2H_2O\right) = 126$

Hence, for preparing 1000 ml of 1 M oxalic acid, weight of oxalic acid crystals required = 126 g

∴ For preparing 250 ml of 0.1 M solution,

$$\text{oxalic acid crystals required} = \frac{126}{1000} \times 250 \times 0.1 = 3.150 \text{ g.}$$

APPARATUS

Watch glass, analytical balance, weight box, fractional weight box, 250 ml beaker, glass rod, 250 ml measuring flask and wash bottle.

CHEMICALS REQUIRED

Oxalic acid crystals and distilled water.

PROCEDURE

1. Take a watch glass, wash it with distilled water and then dry it.
2. Weigh the clean and dried watch glass accurately and record its weight in the notebook.
3. Weigh 3.150 g oxalic acid on the watch glass accurately and record this weight in the notebook.
4. Transfer gently and carefully the oxalic acid from the watch glass into a clean 250 ml measuring flask using a funnel. Wash the watch glass with distilled water with the help of a wash bottle to transfer the particles sticking to it into the funnel [Fig. 11.15].

   The volume of distilled water for this purpose should not be more than 50 ml.
5. Finally wash the funnel well with distilled water with the help of a wash bottle to transfer the solution sticking to the funnel into the measuring flask [Fig. 11.16].
6. Swirl the measuring flask till solid oxalic acid dissolves.
7. Add enough distilled water to the measuring flask carefully, upto just below the etched mark on it, with the help of a wash bottle.

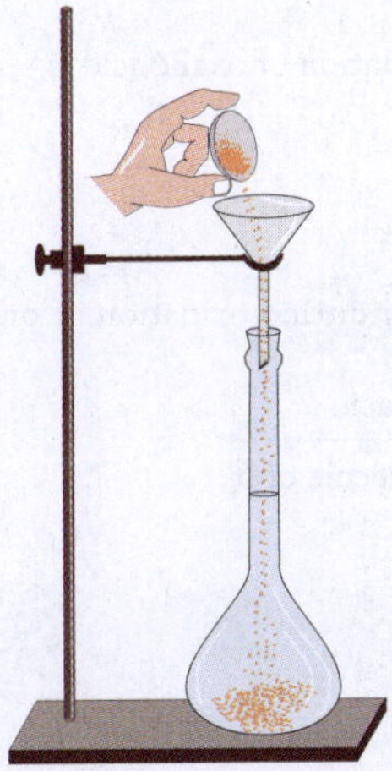

Fig. 11.15. Transferring oxalic acid to the flask.

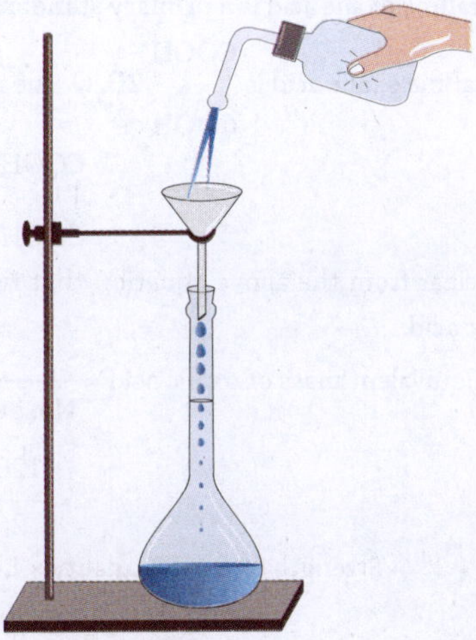

Fig. 11.16. Adding water.

8. Add the last few drops of distilled water with a pipette or a dropper until the lower level of the meniscus just touches the mark on the measuring flask [Fig. 11.17].
9. Stopper the measuring flask and shake gently to make the solution uniform throughout. **Label it as 0.1 M oxalic acid solution [Fig. 11.18].**

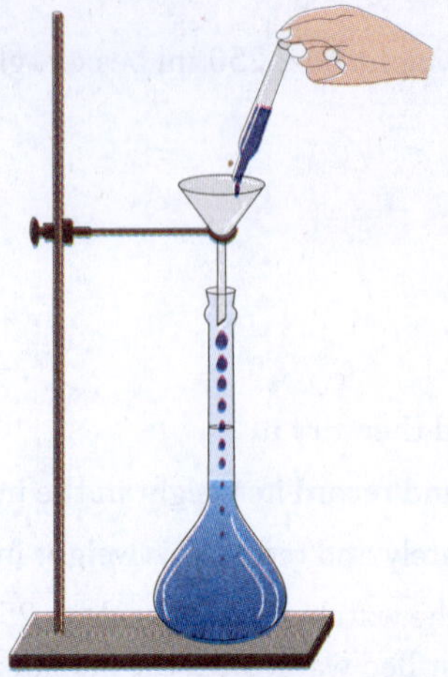

Fig. 11.17. Adding last small amount of water dropwise.

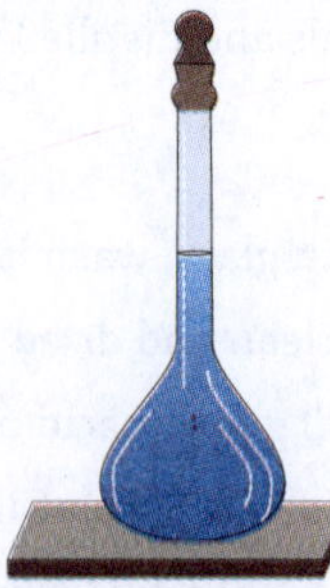

Fig. 11.18. Standard solution of oxalic acid.

KNOWLEDGE PLUS

AIM: ***Prepare 250 ml of a 0.1 N solution of oxalic acid from crystalline oxalic acid.***

THEORY

Crystalline oxalic acid is a primary standard, its standard solution can be prepared directly. The formula for crystalline oxalic acid is $\begin{array}{c}COOH\\|\\COOH\end{array} \cdot 2H_2O$. The ionic equation for the oxidation of oxalic acid is

$$\begin{array}{c}COOH\\|\\COOH\end{array} \longrightarrow 2CO_2 + 2H^+ + 2e^-$$

It is clear from the above equation that two electrons are given out during oxidation of one molecule of oxalic acid.

$$\therefore \quad \text{Equivalent mass of oxalic acid} = \frac{\text{Molecular mass of oxalic acid}}{\text{No. of electrons lost by one molecule of it}}$$

$$= \frac{126}{2} = 63$$

$$\text{Strength } (g/l) = \text{Normality} \times \text{Equivalent mass}$$

$$= \frac{1}{10} \times 63 = 6.3 \text{ g/l}$$

$\therefore$ For preparing 1 litre of $\frac{N}{10}$ oxalic acid solution 6.3 g of it have to be dissolved.

$\therefore$ For preparing 250 ml of $\frac{N}{10}$ oxalic acid, oxalic acid crystals required

$$= \frac{6.3}{1000} \times 250 = 1.575 \text{ g.}$$

Procedure and other details are same as in Experiment 11.1.

Note: *In case the meniscus becomes higher than the mark due to carelessness, reject the solution as the strength of this solution cannot be known, since the volume of the solution is not definite. Start afresh.*

EXPERIMENT 11.2

AIM: ***Preparation of 250 ml of 0.05 M solution of Mohr's salt.***

THEORY

Molecular mass of Mohr's salt, $FeSO_4.(NH_4)_2SO_4.6H_2O = 392$

Hence, for preparing 1000 ml of 1 M Mohr's salt solution,

Mohr's salt required = 392 g

$\therefore$ For preparing 250 ml of $\frac{M}{20}$ Mohr's salt solution,

$$\text{Mohr's salt required} = \frac{392}{1000} \times 250 \times \frac{1}{20} = 4.9 \text{ g}$$

APPARATUS

Watch glass, weight box, fractional weight box, 250 ml beaker, glass rod, 250 ml measuring flask dropper and wash bottle.

CHEMICALS REQUIRED

Mohr's salt, conc. H_2SO_4 and distilled water.

PROCEDURE

1. Weigh the clean and dry watch glass and record its weight in the notebook.
2. Weigh accurately 4.9 g of Mohr's salt crystals on the watch glass and record the weight in the notebook.
3. Transfer carefully the weighed Mohr's salt from the watch glass into a clean 250 ml measuring flask using a funnel.
4. Wash the watch glass thoroughly with distilled water to transfer the sticking salt completely into the flask. Dissolve the salt in the beaker with gentle stirring.
5. Add about 5 ml of conc. H_2SO_4 to the solution in the measuring flask to check the hydrolysis of ferrous sulphate.
6. Wash the funnel with distilled water and transfer the washings into the measuring flask.
7. Add enough distilled water to the measuring flask carefully upto just below the etched mark on its neck with the help of wash bottle.
8. *Add the last few drops of distilled water with a pipette or a dropper until the lower level of the meniscus just touches the mark on the measuring flask.*
9. Stopper the measuring flask and shake it gently to make the solution homogeneous (*i.e.*, uniform throughout) and **label it as $\frac{M}{20}$ Mohr's salt solution.**

KNOWLEDGE PLUS

AIM: ***Prepare 250 ml of 0.05 N solution of Mohr's salt.***

THEORY

The molecular formula of Mohr's salt is $FeSO_4.(NH_4)_2SO_4.6H_2O$.

The ionic equation for the oxidation of Mohr's salt is

$$Fe^{2+} \longrightarrow Fe^{3+} + e^-$$

It is clear from the above equation that one electron is given out during the oxidation of one molecule of Mohr's salt.

$$\therefore \quad \text{Equivalent mass of Mohr's salt} = \frac{\text{Molecular mass of Mohr's salt}}{\text{Number of electrons lost by one molecule of it}}$$

$$= \frac{392}{1} = 392$$

$$\text{Strength (g/litre)} = \text{Normality} \times \text{Equivalent mass}$$

$$= \frac{1}{20} \times 392 = 19.6 \text{ g/litre}$$

For preparing 250 ml of $\frac{N}{20}$ Mohr's salt solution Mohr's salt needed

$$= \frac{19.6}{1000} \times 250 = 4.9 \text{ g}$$

Procedure and other details are same as in Exp. 11.2.

11.07 EXPERIMENTS ON POTASSIUM PERMANGANATE TITRATIONS—PERMANGANOMETRIC TITRATIONS

Potassium permanganate is a strong oxidizing agent in the presence of sulphuric acid.

$$2KMnO_4 + 3H_2SO_4 \longrightarrow K_2SO_4 + 2MnSO_4 + 3H_2O + 5[O]$$

Potassium permanganate can be used for the estimation of oxalic acid (or oxalates) and ferrous salts. It oxidizes oxalic acid to CO_2 and ferrous salts to ferric salts in acidic medium.

$$\begin{matrix} COOH \\ | \\ COOH \end{matrix} + [O] \longrightarrow 2CO_2 + H_2O$$

$$2FeSO_4 + H_2SO_4 + [O] \longrightarrow Fe_2(SO_4)_3 + H_2O$$

In $KMnO_4$ titrations, no external indicator is required. ***$KMnO_4$ acts as self indicator.*** The end point is appearance of permanent pinkish tinge with the last drop of $KMnO_4$ solution.

CHEMICAL EQUATIONS

For oxidation of oxalic acid with $KMnO_4$ solution

Molecular equations

$$2KMnO_4 + 3H_2SO_4 \longrightarrow K_2SO_4 + 2MnSO_4 + 3H_2O + 5[O]$$

$$\begin{matrix} COOH \\ | \\ COOH \end{matrix} \cdot 2H_2O + [O] \xrightarrow{60\text{–}70^\circ C} 2CO_2 + 3H_2O] \times 5$$

$$2KMnO_4 + 3H_2SO_4 + 5 \begin{matrix} COOH \\ | \\ COOH \end{matrix} \cdot 2H_2O \longrightarrow K_2SO_4 + 2MnSO_4 + 18H_2O + 10CO_2$$

Ionic equations

$$MnO_4^- + 8H^+ + 5e^- \longrightarrow Mn^{2+} + 4H_2O] \times 2$$

$$C_2O_4^{2-} \longrightarrow 2CO_2 + 2e^-] \times 5$$

$$2MnO_4^- + 16H^+ + 5C_2O_4^{2-} \longrightarrow 2Mn^{2+} + 8H_2O + 10CO_2$$

For oxidation of Mohr's salt solution or Fe^{2+} ions with $KMnO_4$ solution

Molecular equations

$$FeSO_4(NH_4)_2\,SO_4.6H_2O \longrightarrow FeSO_4 + (NH_4)_2SO_4 + 6H_2O] \times 10$$

$$2KMnO_4 + 3H_2SO_4 \longrightarrow K_2SO_4 + 2MnSO_4 + 3H_2O + 5[O]$$

$$2FeSO_4 + H_2SO_4 + [O] \longrightarrow Fe_2(SO_4)_3 + H_2O] \times 5$$

$$2KMnO_4 + 8H_2SO_4 + 10FeSO_4\,(NH_4)_2SO_4.6H_2O \longrightarrow K_2SO_4 + 2MnSO_4 + 5Fe_2\,(SO_4)_3 + 10(NH_4)_2SO_4 + 68H_2O$$

Ionic equations

$$MnO_4^- + 8H^+ + 5e^- \longrightarrow Mn^{2+} + 4H_2O$$

$$Fe^{2+} \longrightarrow Fe^{3+} + e^-] \times 5$$

$$MnO_4^- + 8H^+ + 5Fe^{2+} \longrightarrow 5Fe^{3+} + Mn^{2+} + 4H_2O$$

PROCESS OF TITRATION

The process of titration is carried out as under:

1. Support a cleaned and rinsed burette with a burette clamp. Close the stopcock and, with the help of a funnel, fill the burette with $KMnO_4$ solution to just above the zero mark. Open the stop cock briefly to remove any air bubbles in the tip.

   Note: The burette used in permanganate titrations must have a glass stop cock as rubber is attacked by permanganate solution.

2. Take a pipette and wash it with water. Rinse the pipette with the solution to be pipetted out. Pipette out 20.0 ml of the solution to be titrated in a washed titration flask.
3. Place a glazed white tile below the burette and place the titration flask on the glazed tile below the burette nozzle. Adjust the height of the burette so that the nozzle tip just enters the mouth of the titration flask.
4. Note the initial reading of the burette and run out the solution from the burette (one ml at a time). During titration, operate the stopcock with your left hand and constantly swirl the flask with the right hand. (See Fig. 11.19).
5. Continue running more of the solution from the burette into the titration flask. The solution should fall directly into the solution of titration flask. It should not fall on the walls of flask.

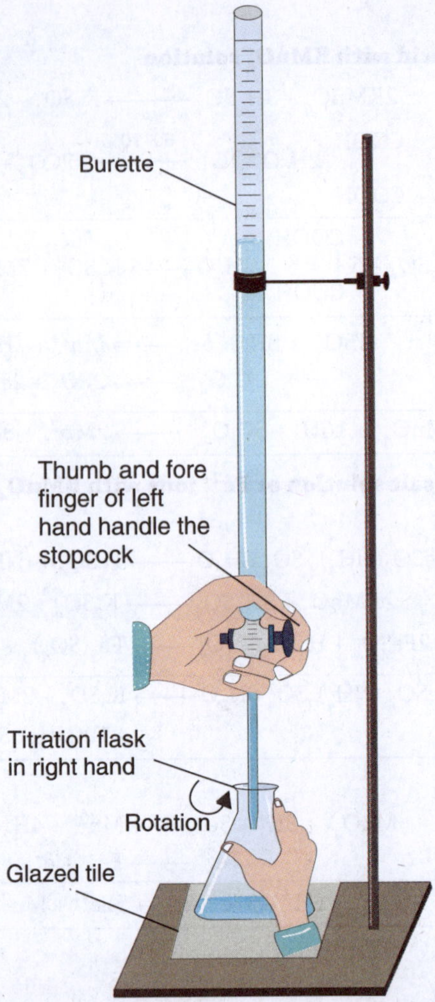

Fig. 11.19. Correct way of handling a burette.

6. Stop addition of the solution when the end point is reached. The end point is indicated by the appearance of permanent light pink due to slight excess of permanganate solution. Take final reading of the burette. *The difference between the final and the initial readings gives rough volume of the solution used for completion of the reaction.*
7. The solution from the titration flask is thrown in the sink and the titration flask is washed thoroughly first by keeping it under tap water and then with a little of distilled water. ***Do not rinse the titration flask.***
8. Pour more solution in the burette.
9. Pipette out 20.0 ml of the solution into the titration flask.
10. Note the initial reading of the burette. Run solution from the burette into the titration flask slowly with constant shaking. Continue running of the solution till the volume added is 1 ml less than the rough volume found out in the first titration. Now add solution from the burette dropwise. (Add a drop, close the stop cock, shake and find out if the end point has been attained).

11. *Continue adding solution dropwise from the burette, till by addition of last single drop, the end point is attained.*
12. Note down the final reading of the burette. The difference between the final and initial readings of the burette gives the exact volume of the solution required for completion of the reaction.
13. *Check the correctness of the end point by adding one drop of solution (taken in the titration flask) with the help of a pipette*. Restoration of original colour confirms the correctness of the end point.
14. Perform 5–6 titrations so that at least three concordant readings (difference not more than 0.05 ml) are obtained.

Important Instructions for $KMnO_4$ Titrations

1. $KMnO_4$ solution is always taken in the burette.
2. Avoid the use of a burette having a rubber tap as $KMnO_4$ attacks rubber.
3. Add about an equal volume of dil. H_2SO_4 (~ 2M) to the solution to be titrated (say a full test tube for 20 ml of the solution) before adding $KMnO_4$.

   HCl cannot be used as it gets oxidized to Cl_2 by $KMnO_4$. HNO_3 also can not be used as it itself is a strong oxidizing agent.
4. If oxalic acid or some oxalate is to be titrated, add required amount of dil. H_2SO_4 and heat the flask to 60°—70°C on a wire gauge. In order to get some idea about the temperature of the solution touch the flask to the back side of your hand (Fig. 11.20). When it becomes just unbearable to touch, the required temperature is reached. The purpose of heating is to increase the rate of reaction which otherwise is slow at room temperature.
5. In case of ferrous salts, no warming is required.
6. Read the upper meniscus while taking burette reading with $KMnO_4$ solution.
7. In case, on addition of $KMnO_4$ a brown ppt. appears, this shows that either H_2SO_4 has not been added or has been added in insufficient amount. In such a case, throw away the solution and titrate again.

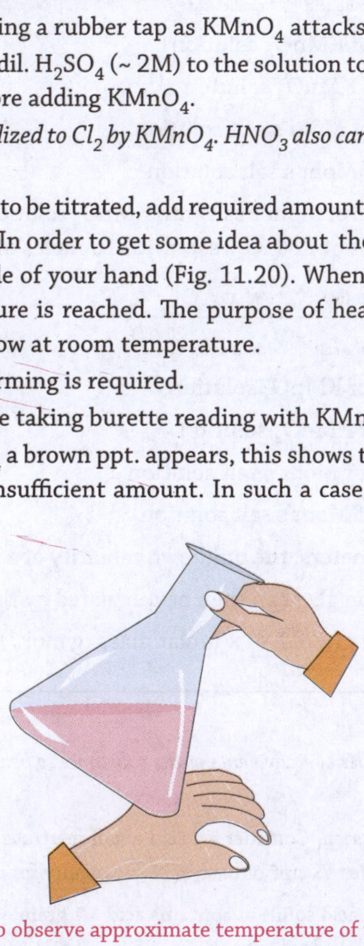

Fig. 11.20. How to observe approximate temperature of the solution.

8. Potassium permanganate does not dissolve into water readily. It is dissolved by the process of extraction. Transfer the weighed $KMnO_4$ into a beaker and add into it 20–30 ml of distilled water and stir. Transfer the solution into a measuring flask. Add more distilled water (20–30 ml) into the beaker and repeat the operation till the permanganate completely dissolves. Add more distilled water into the measuring flask *till the lower meniscus of the solution is in line with the mark on the neck.*

   Stopper the measuring flask and shake to get the solution of uniform strength.

CALCULATIONS

The strength of the unknown solution in terms of molarity may be determined by the following equation:

$$\frac{\text{Molarity } (M_1) \times \text{Volume } (V_1)}{\text{Molarity } (M_2) \times \text{Volume } (V_2)} = \frac{\text{Stoichiometric coefficient of the reactant 1}}{\text{Stoichiometric coefficient of the reactant 2}}$$

For example, for oxidation of oxalic acid with $KMnO_4$,

From the overall balanced chemical equation it is clear that 2 moles of $KMnO_4$ react with 5 moles of oxalic acid.

$$\therefore \quad \frac{M_{KMnO_4} \times V_{KMnO_4}}{M_{\text{oxalic acid}} \times V_{\text{oxalic acid}}} = \frac{2}{5}$$

where, M_{KMnO_4} = Molarity of $KMnO_4$ solution

V_{KMnO_4} = Volume of $KMnO_4$ solution

$M_{\text{oxalic acid}}$ = Molarity of Mohr's salt solution

$V_{\text{oxalic acid}}$ = Volume of Mohr's salt solution

Similarly, for the oxidation of Mohr's salt with $KMnO_4$ solution, from the overall balanced chemical equation, it is clear that 2 moles of $KMnO_4$ react with 10 moles of Mohr's salt.

$$\therefore \quad \frac{M_{KMnO_4} \times V_{KMnO_4}}{M_{\text{Mohr's salt}} \times V_{\text{Mohr's salt}}} = \frac{2}{10}$$

where, M_{KMnO_4} = Molarity of $KMnO_4$ solution

V_{KMnO_4} = Volume of $KMnO_4$ solution

$M_{\text{Mohr's salt}}$ = Molarity of Mohr's salt solution

$V_{\text{Mohr's salt}}$ = Volume of Mohr's salt solution

Knowing any three parameters, the unknown molarity of a solution can be calculated.

Strength of the solution in gL^{-1} can then be calculated by the following relation:

Strength (gL^{-1}) = Molarity (mol L^{-1}) × Molar mass (g mol^{-1}).

KNOWLEDGE PLUS

According to this law, *the number of equivalents of the substance to be titrated (titre) is equal to the number of equivalents of the titrant used.*

Derivation of the normality equation. Consider an acid alkali neutralization reaction. Let V_1 cm^3 of an acid solution of N_1 normality requires V_2 cm^3 of base of N_2 normality for complete neutralization.

We know that 1000 cm^3 of 1 N acid solution contains acid = 1 gram equivalent.

V_1 cm^3 of N_1 acid contains acid = $\frac{1}{1000} \times V_1 \times N_1$ gm equivalents. Thus, number of gram equivalents of acid in V_1 cm^3 solution = $\frac{V_1N_1}{1000}$.

Similarly, number of gram equivalents of base in V_2 cm^3 of its N_2 solution = $\frac{V_2N_2}{1000}$.

By the law of equivalents, at the end point, $\frac{V_1N_1}{1000} = \frac{V_2N_2}{1000}$.

$$\boxed{N_1V_1 = N_2V_2}$$

It is known as **normality equation.** If three factors (V_1, V_2, N_1) are known, the fourth (N_2) can be calculated by using above formula.

Recording of Volumetric Analysis in the Practical Notebook

Left hand page *(with pencil)*	***Right hand page*** *(with ball pen)*
Date Experiment Chemical equation Indicator End point Observations Calculations	Date Experiment Requirement Theory Procedure General calculations

EXPERIMENT 11.3

AIM: ***Prepare 0.05 M solution of ferrous ammonium sulphate (Mohr's salt). Using this solution find out the molarity and strength of the given $KMnO_4$ solution.***

CHEMICAL EQUATIONS

Molecular equations

$$2KMnO_4 + 3H_2SO_4 \longrightarrow K_2SO_4 + 2MnSO_4 + 3H_2O + 5[O]$$

$$[2FeSO_4(NH_4)_2SO_4.6H_2O + H_2SO_4 + [O] \longrightarrow Fe_2(SO_4)_3 + 2(NH_4)_2SO_4 + 13H_2O] \times 5$$

$$2KMnO_4 + 8H_2SO_4 + 10FeSO_4(NH_4)_2SO_4.6H_2O \longrightarrow K_2SO_4 + 2MnSO_4 + 5Fe_2(SO_4)_3 + 10(NH_4)_2SO_4 + 68H_2O$$

Ionic equations

$$MnO_4^- + 8H^+ + 5e^- \longrightarrow Mn^{2+} + 4H_2O$$

$$[Fe^{2+} \longrightarrow Fe^{3+} + e^-] \times 5$$

$$MnO_4^- + 8H^+ + 5Fe^{2+} \longrightarrow 5Fe^{3+} + Mn^{2+} + 4H_2O$$

INDICATOR

$KMnO_4$ is a self-indicator.

END POINT

Colourless to permanent pink colour (*$KMnO_4$ in burette*).

PROCEDURE

1. Prepare 250 ml of 0.05 M Mohr's salt solution by dissolving 4.9 g of Mohr's salt in water as described in experiment 11.3. Rinse the pipette with the 0.05 M Mohr's salt solution and pipette out 20.0 ml of it in a washed titration flask.

2. Rinse and fill the burette with the given $KMnO_4$ solution.
3. Add one test tube (~ 20 ml) full of dilute sulphuric acid (~ 2 M) to the solution in titration flask.
4. Note the initial reading of the burette.
5. Now add $KMnO_4$ solution from the burette till a permanent light pink colour is imparted to the solution in the titration flask on addition of last single drop of $KMnO_4$ solution.
6. Note the final reading of the burette.
7. Repeat the above steps 4–5 times to get a set of three concordant readings.

OBSERVATIONS

Weight of watch glass = ... g

Weight of watch glass + Mohr's salt = ... g

Weight of Mohr's salt = 4.90 g

Volume of Mohr's salt solution prepared = 250 ml

Molarity of Mohr's salt solution = 0.05 M

Volume of Mohr's salt solution taken for each titration = 20.0 ml

S. No.	***Burette Readings***		***Volume of the $KMnO_4$ solution used***
	Initial	***Final***	
1.	—	—	— ml
2.	—	—	— ml
3.	—	—	— ml
4.	—	—	— ml

Concordant volume = x ml (say)

CALCULATIONS

(*a*) Molarity of the $KMnO_4$ solution

From the overall balanced chemical equation, it is clear that 2 moles of $KMnO_4$ react with 10 moles of Mohr's salt.

$$\therefore \qquad \frac{M_{KMnO_4} \times V_{KMnO_4}}{M_{\text{Mohr's salt}} \times V_{\text{Mohr's salt}}} = \frac{2}{10}$$

where, M_{KMnO_4} = Molarity of $KMnO_4$ solution

V_{KMnO_4} = Volume of $KMnO_4$ solution

$M_{\text{Mohr's salt}}$ = Molarity of Mohr's salt solution

$V_{\text{Mohr's salt}}$ = Volume of Mohr's salt solution

$$\frac{M_{KMnO_4} \times x}{0.05 \times 20} = \frac{2}{10}$$

$$M_{KMnO_4} = \frac{2}{10} \times \frac{1}{x} = \frac{2}{10x}$$

(*b*) Strength of the $KMnO_4$ solution

Strength (in g/L) = Molarity × Molar mass

$$= \frac{2}{10x} \times 158 \text{ g/l.}$$

Instructions for the Preparation of Solutions:

Provide the following:

1. Crystals of Mohr's salt
2. M/100 $KMnO_4$ solution (1.58 g/litre)
3. 4 N H_2SO_4.

EXPERIMENT 11.4

AIM: ***Prepare a solution of ferrous ammonium sulphate (Mohr's salt) containing exactly 17.0 g of the salt in one litre. With the help of this solution, determine the molarity and the concentration of $KMnO_4$ in the given solution.***

CHEMICAL EQUATIONS

Molecular equations

$$2KMnO_4 + 3H_2SO_4 \longrightarrow K_2SO_4 + 2MnSO_4 + 3H_2O + 5[O]$$

$$[2FeSO_4(NH_4)_2SO_4.6H_2O + H_2SO_4 + [O] \longrightarrow Fe_2(SO_4)_3 + 2(NH_4)_2SO_4 + 13H_2O] \times 5$$

$$2KMnO_4 + 8H_2SO_4 + 10FeSO_4(NH_4)_2SO_4.6H_2O \longrightarrow K_2SO_4 + 2MnSO_4 + 5Fe_2(SO_4)_3 + 10(NH_4)_2SO_4 + 68H_2O$$

Ionic equations

$$MnO_4^- + 8H^+ + 5e^- \longrightarrow Mn^{2+} + 4H_2O$$

$$[Fe^{2+} \longrightarrow Fe^{3+} + e^-] \times 5$$

$$MnO_4^- + 8H^+ + 5Fe^{2+} \longrightarrow 5Fe^{3+} + Mn^{2+} + 4H_2O$$

INDICATOR

$KMnO_4$ is a self-indicator.

END POINT

Colourless to permanent pink colour (*$KMnO_4$ in burette*).

PROCEDURE

1. Weigh exactly 4.250 g of Mohr's salt on a watch glass and dissolve in water to prepare exactly 250 ml of solution with the help of a 250 ml measuring flask. Rinse the pipette with the prepared Mohr's salt solution and pipette out 20.0 ml of it in a washed titration flask.
2. Rinse and fill the burette with the given $KMnO_4$ solution.
3. Add one test tube (~ 20 ml) full of dilute sulphuric acid (~ 2 M) to the solution in titration flask.
4. Note the initial reading of the burette.
5. Now add $KMnO_4$ solution from the burette till a permanent light pink colour is imparted to the solution in the titration flask on addition of a last single drop of $KMnO_4$ solution.
6. Note the final reading of the burette.
7. Repeat the above steps 4–5 times to get three concordant readings.

OBSERVATIONS

Weight of watch glass = ... g

Weight of watch glass + Mohr's salt = ... g

Weight of Mohr's salt = 4.250 g

Volume of Mohr's salt solution prepared = 250 ml

Volume of Mohr's salt solution taken for each titration = 20.0 ml

S. No.	***Burette Readings***		***Volume of the $KMnO_4$ solution used***
	Initial	***Final***	
1.	—	—	— ml
2.	—	—	— ml
3.	—	—	— ml
4.	—	—	— ml

Concordant volume = *x* ml (say)

CALCULATIONS

Concentration of Mohr's salt, ferrous ammonium sulphate, $FeSO_4.(NH_4)_2SO_4.6H_2O$ in the prepared solution = 17.0 g/litre.

Molecular mass of Mohr's salt, $FeSO_4.(NH_4)_2SO_4.6H_2O$ = 392

$$\text{Molarity of Mohr's salt solution} = \frac{\text{Strength (g/litre)}}{\text{Molar mass}} = \frac{17.0}{392}$$

Calculation of molarity of $KMnO_4$ solution

From the overall balanced chemical equation, it is clear that 2 moles of $KMnO_4$ react with 10 moles of Mohr's salt.

$$\therefore \qquad \frac{M_{KMnO_4} \times V_{KMnO_4}}{M_{\text{Mohr's salt}} \times V_{\text{Mohr's salt}}} = \frac{2}{10}$$

where, M_{KMnO_4} = Molarity of $KMnO_4$ solution

V_{KMnO_4} = Volume of $KMnO_4$ solution

$M_{\text{Mohr's salt}}$ = Molarity of Mohr's salt solution

$V_{\text{Mohr's salt}}$ = Volume of Mohr's salt solution

$$\frac{M_{KMnO_4} \times x}{17/392 \times 20} = \frac{2}{10}$$

From this equation, molarity of $KMnO_4$ solution can be calculated.

Calculation of strength of $KMnO_4$ solution

Strength (in g/litre) = Molarity × Molar mass

$= M_{KMnO_4} \times 158$ g/l.

> **Instructions for the Preparation of Solutions:**
>
> Provide the following:
>
> 1. Mohr's salt
> 2. M/100 $KMnO_4$ solution (1.58 g/litre)
> 3. 4 N H_2SO_4.

EXPERIMENT 11.5

AIM: ***Prepare 0.05 M ferrous ammonium sulphate (Mohr's salt) solution. Find out the percentage purity of impure $KMnO_4$ sample 2.0 g of which have been dissolved per litre.***

CHEMICAL EQUATIONS

Molecular equations

$$2KMnO_4 + 3H_2SO_4 \longrightarrow K_2SO_4 + 2MnSO_4 + 3H_2O + 5[O]$$

$$2FeSO_4(NH_4)_2SO_4.6H_2O + H_2SO_4 + [O] \longrightarrow Fe_2(SO_4)_3 + 2(NH_4)_2SO_4 + 13H_2O] \times 5$$

$$2KMnO_4 + 8H_2SO_4 + 10FeSO_4(NH_4)_2SO_4.6H_2O \longrightarrow K_2SO_4 + 2MnSO_4 + 5Fe_2(SO_4)_3 + 10(NH_4)_2SO_4 + 68H_2O$$

Ionic equations

$$MnO_4^- + 8H^+ + 5e^- \longrightarrow Mn^{2+} + 4H_2O$$

$$Fe^{2+} \longrightarrow Fe^{3+} + e^-] \times 5$$

$$MnO_4^- + 8H^+ + 5Fe^{2+} \longrightarrow 5Fe^{3+} + Mn^{2+} + 4H_2O$$

INDICATOR

$KMnO_4$ is a self-indicator.

END POINT

Colourless to permanent pink colour (*$KMnO_4$ in burette*).

PROCEDURE

1. Prepare 250 ml of 0.05 M Mohr's salt solution by dissolving 4.9 g of Mohr's salt in water (as described in experiment 11.4). Rinse the pipette with the 0.05 M Mohr's salt solution and pipette out 20.0 ml of it in a washed titration flask.
2. Rinse and fill the burette with the given $KMnO_4$ solution.
3. Add one test-tube (~ 20 ml) full of dilute sulphuric acid (~ 2 M) to the solution in titration flask.
4. Note the initial reading of the burette.
5. Now add $KMnO_4$ solution from the burette till a permanent light pink colour is imparted to the solution in the titration flask on addition of a last single drop of $KMnO_4$ solution.
6. Note the final reading of the burette.
7. Repeat the above steps 4–5 times to get three concordant readings.

OBSERVATIONS

Weight of watch glass = ... g

Weight of watch glass + Mohr's salt = ... g

Weight of Mohr's salt = 4.9 g

Volume of solution prepared = 250 ml

Solution taken in burette = $KMnO_4$ solution

Volume of Mohr's salt solution taken for each titration = 20.0 ml

S. No.	Burette Readings		Volume of the $KMnO_4$ solution used
	Initial	Final	
1.	—	—	— ml
2.	—	—	— ml
3.	—	—	— ml
4.	—	—	— ml

Concordant volume = x ml (say).

CALCULATIONS

Calculation of molarity of $KMnO_4$ solution

From the overall balanced chemical equation, it is clear that 2 moles of $KMnO_4$ react with 10 moles of Mohr's salt.

$$\therefore \quad \frac{M_{KMnO_4} \times V_{KMnO_4}}{M_{Mohr's\ salt} \times V_{Mohr's\ salt}} = \frac{2}{10}$$

where, M_{KMnO_4} = Molarity of $KMnO_4$ solution

V_{KMnO_4} = Volume of $KMnO_4$ solution

$M_{Mohr's\ salt}$ = Molarity of Mohr's salt solution

$V_{Mohr's\ salt}$ = Volume of Mohr's salt solution

$$\frac{M_{KMnO_4} \times x}{0.05 \times 20} = \frac{2}{10}$$

$$M_{KMnO_4} = \frac{2}{10} \times \frac{1}{x} = \frac{2}{10x}$$

Calculation of strength of $KMnO_4$ solution

Strength (in g/litre) = Molarity × Molar mass of $KMnO_4$

$$= M_{KMnO_4} \times 158 \text{ g/l}$$

$$= \frac{2}{10x} \times 158 = a \text{ g/l (say).}$$

Calculation of percentage purity of the given sample

Percentage purity $= \frac{a}{2} \times 100$

Instructions for the Preparation of Solutions

Provide the following:

1. Crystals of Mohr's salt
2. $KMnO_4$ solution (1.6 g/litre)
3. 2M H_2SO_4.

EXPERIMENT 11.6

AIM: ***Determine the number of molecules of water of crystallisation in a sample of Mohr's salt, $FeSO_4(NH_4)_2SO_4.nH_2O$. Provided 0.01 M $KMnO_4$.***

CHEMICAL EQUATIONS

Molecular equations

$$FeSO_4(NH_4)_2SO_4.\,nH_2O \longrightarrow FeSO_4 + (NH_4)_2SO_4 + nH_2O$$

$$2KMnO_4 + 3H_2SO_4 \longrightarrow K_2SO_4 + 2MnSO_4 + 3H_2O + 5[O]$$

$$2FeSO_4 + H_2SO_4 + [O] \longrightarrow Fe_2(SO_4)_3 + H_2O] \times 5$$

Ionic equations

$$MnO_4^- + 8H^+ + 5e^- \longrightarrow Mn^{2+} + 4H_2O$$

$$Fe^{2+} \longrightarrow Fe^{3+} + e^-] \times 5$$

$$MnO_4^- + 8H^+ + 5Fe^{2+} \longrightarrow 5Fe^{3+} + Mn^{2+} + 4H_2O$$

THEORY

Prepare a solution of Mohr's salt with known strength (g/litre). Molarity of ferrous ammonium sulphate can be determined by directly titrating it against standard 0.01 M $KMnO_4$ solution.

Molecular mass = strength/molarity.

Substituting the value of strength and value of molarity as calculated above, the molecular mass of Mohr's salt can be calculated. Suppose it comes out to be M.

Theoretical molecular mass of Mohr's salt, $FeSO_4.(NH_4)_2SO_4.\,nH_2O$

$$= 284 + 18n$$

$$\therefore \quad 284 + 18n = M$$

whence, n can be calculated.

INDICATOR

$KMnO_4$ is a self-indicator.

END POINT

Colourless to permanent pink (*$KMnO_4$ in burette*).

PROCEDURE

1. Weigh exactly 4.90 g of Mohr's salt and dissolve in water to prepare exactly 250 ml of solution, using a 250 ml measuring flask. Rinse the pipette with the prepared Mohr's salt solution and pipette out 20.0 ml of it in a washed titration flask.
2. Rinse and fill the burette with 0.01 M $KMnO_4$ solution.
3. Add one test tube (~ 20 ml) full of dilute sulphuric acid (~ 2 M) to the solution in the titration flask.

4. Note the initial reading of the burette.
5. Add $KMnO_4$ solution into the titration flask from the burette till a permanent light pink colour is imparted to the solution in the titration flask on addition of a last single drop of $KMnO_4$ solution.
6. Note the final reading of the burette.
7. Repeat the above steps 4–5 times to get three concordant readings.

OBSERVATIONS

Weight of watch glass = ... g

Weight of watch glass + Mohr's salt = ... g

Weight of Mohr's salt = 4.90 g

Volume of Mohr's salt solution prepared = 250 ml

Volume of Mohr's salt solution taken for each titration = 20.0 ml

Molarity of $KMnO_4$ solution = 0.01 M

S. No.	*Burette Readings*		*Volume of the $KMnO_4$ solution used*
	Initial	*Final*	
1.	—	—	— ml
2.	—	—	— ml
3.	—	—	— ml
4.	—	—	— ml

Concordant volume = x ml (say)

CALCULATIONS

Molarity of the standard $KMnO_4$ solution = 0.01 M

x ml of 0.01 M $KMnO_4$ react with 20.0 ml of the given Mohr's salt solution.

From the chemical equations, it is clear that one mole of $KMnO_4$ reacts with 5 moles Mohr's salt.

$$\therefore \quad \frac{M_{\text{Mohr's salt}} \times V_{\text{Mohr's salt}}}{M_{KMnO_4} \times V_{KMnO_4}} = \frac{5}{1}$$

$$\frac{M_{\text{Mohr's salt}} \times 20.0}{0.01 \times x} = \frac{5}{1}$$

$$M_{\text{Mohr's salt}} = \frac{5 \times x \times 0.01}{20} = \frac{x}{400}$$

$$\text{Molecular mass of Mohr's salt} = \frac{\text{Strength in g/l}}{\text{Molarity}}$$

$$= \frac{19.6}{x/400} = \text{M (say)}$$

Calculation of no. of molecules of water of crystallisation

Theoretically, molecular mass of Mohr's salt, $FeSO_4.(NH_4)_2SO_4.nH_2O$

$$= 284 + 18n$$

Equating this with the experimentally determined molecular mass, we get

$$M = 284 + 18n$$

$$n = \frac{M - 284}{18}$$

Instructions for the Preparation of Solutions

Provide the following:

1. Mohr's salt
2. $KMnO_4$ solution (1.58 g/litre)
3. 2M H_2SO_4.

EXPERIMENT 11.7

AIM: ***Prepare $\frac{M}{40}$ solution of oxalic acid. With its help, determine the molarity and strength of the given solution of potassium permanganate ($KMnO_4$).***

CHEMICAL EQUATIONS

Molecular equations

$$2KMnO_4 + 3H_2SO_4 \longrightarrow K_2SO_4 + 2MnSO_4 + 3H_2O + 5[O]$$

$$\left.\begin{matrix}COOH\\|\\COOH\end{matrix}\right. . 2H_2O + [O] \xrightarrow{60-70°C} 2CO_2 + 3H_2O] \times 5$$

$$2KMnO_4 + 2H_2SO_4 + 5\begin{matrix}COOH\\|\\COOH\end{matrix} . 2H_2O \longrightarrow K_2SO_4 + 2MnSO_4 + 18H_2O + 10CO_2$$

Ionic equations

$$MnO_4^- + 8H^+ + 5e^- \longrightarrow Mn^{2+} + 4H_2O] \times 2$$

$$C_2O_4^{2-} \longrightarrow 2CO_2 + 2e^-] \times 5$$

$$2MnO_4^- + 16H^+ + 5C_2O_4^{2-} \longrightarrow 2Mn^{2+} + 8H_2O + 10CO_2$$

INDICATOR

$KMnO_4$ is a self-indicator.

END POINT

Colourless to permanent pink colour (*$KMnO_4$ in burette*).

PROCEDURE

1. Weigh 1.580 g of oxalic acid crystals and dissove them in water to prepare 500 ml of $\frac{M}{40}$ oxalic acid solution using a 500 ml measuring flask. Rinse the pipette with the $\frac{M}{40}$ oxalic acid solution and pipette out 20 ml of it in a washed titration flask.
2. Rinse and fill the burette with the given $KMnO_4$ solution.
3. Add one test tube (~ 20 ml) full of dilute sulphuric acid (~ 2 M) to the solution in titration flask.
4. Note the initial reading of the burette.
5. Heat the flask to 60–70°C and add $KMnO_4$ solution from the burette till a permanent light pink colour is imparted to the solution in the titration flask on addition of a last single drop of $KMnO_4$ solution.
6. Note the final reading of the burette.
7. Repeat the above steps 4–5 times to get three concordant readings.

OBSERVATIONS

Weight of watch glass = ... g

Weight of (watch glass + oxalic acid) = ... g

Weight of oxalic acid = 1.580 g

Volume of solution prepared = 500 ml

Molarity of oxalic acid solution = M/40

Volume of oxalic acid solution taken for each titration = 20.0 ml.

S. No.	*Burette Readings*		*Volume of the $KMnO_4$ solution used*
	Initial	*Final*	
1.	—	—	— ml
2.	—	—	— ml
3.	—	—	— ml
4.	—	—	— ml

Concordant volume = *x* ml (say)

CALCULATIONS

(a) Calculation of molarity of the $KMnO_4$ solution

From the overall balanced chemical equation it is clear that 2 moles of $KMnO_4$ react with 5 moles of oxalic acid.

$$\therefore \quad \frac{M_{KMnO_4} \times V_{KMnO_4}}{M_{oxalic\ acid} \times V_{oxalic\ acid}} = \frac{2}{5}$$

where, M_{KMnO_4} = Molarity of $KMnO_4$ solution

V_{KMnO_4} = Volume of $KMnO_4$ solution

$M_{oxalic\ acid}$ = Molarity of Mohr's salt solution
$V_{oxalic\ acid}$ = Volume of Mohr's salt solution

$$\frac{M_{KMnO_4} \times x}{1/40 \times 20} = \frac{2}{5}$$

$$M_{KMnO_4} = \frac{1}{5x}$$

(b) Calculation of strength of the $KMnO_4$ solution

$$\text{Strength (in g/l)} = \text{Molarity} \times \text{Molar mass}$$
$$= \frac{1}{5x} \times 158$$

Instructions for the Preparation of Solutions

Provide the following :

1. Oxalic acid crystals
2. $KMnO_4$ solution (1.58 g/litre)
3. 2M H_2SO_4.

EXPERIMENT 11.8

AIM: ***Find out the percentage purity of impure sample of oxalic acid. You are supplied 0.01 M $KMnO_4$ solution.***

CHEMICAL EQUATIONS

Molecular equations

$$2KMnO_4 + 3H_2SO_4 \longrightarrow K_2SO_4 + 2MnSO_4 + 3H_2O + 5[O]$$

$$\begin{matrix} COOH \\ | \\ COOH \end{matrix} . 2H_2O + [O] \xrightarrow{60\text{–}70°C} 2CO_2 + 3H_2O] \times 5$$

$$2KMnO_4 + 2H_2SO_4 + 5 \begin{matrix} COOH \\ | \\ COOH \end{matrix} . 2H_2O \longrightarrow K_2SO_4 + 2MnSO_4 + 18H_2O + 10CO_2$$

Ionic equations

$$MnO_4^- + 8H^+ + 5e^- \longrightarrow Mn^{2+} + 4H_2O] \times 2$$

$$C_2O_4^{2-} \longrightarrow 2CO_2 + 2e^-] \times 5$$

$$2MnO_4^- + 16H^+ + 5C_2O_4^{2-} \longrightarrow 2Mn^{2+} + 8H_2O + 10CO_2$$

INDICATOR

$KMnO_4$ is a self-indicator.

END POINT

Colourless to permanent pink colour (*$KMnO_4$ in burette*).

PROCEDURE

1. Weigh exactly 2.0 g of oxalic acid and dissolve in water to prepare 500 ml of its solution using a 500 ml measuring flask. Rinse the pipette with the oxalic acid solution and pipette out 20 ml of it in a washed titration flask.
2. Rinse and fill the burette with the given 0.01 M $KMnO_4$ solution.
3. Add one test tube (~ 20 ml) full of dilute sulphuric acid (~ 2 M) to the solution in titration flask.
4. Note the initial reading of the burette.
5. Heat the flask to 60–70°C and add $KMnO_4$ solution from the burette till a permanent light pink colour just appears in the solution in the titration flask.
6. Note the final reading of the burette.
7. Repeat the above steps 4–5 times to get three concordant readings.

OBSERVATIONS

Weight of watch glass = ... g

Weight of (watch glass + oxalic acid) = ... g

Weight of oxalic acid = 2.00 g

Volume of oxalic acid solution prepared = 500 ml

Solution taken in burette = 0.01 M $KMnO_4$

Volume of oxalic acid solution taken for each titration = 20.0 ml.

S. No.	*Burette Readings*		*Volume of the $KMnO_4$ solution used*
	Initial	*Final*	
1.	—	—	— ml
2.	—	—	— ml
3.	—	—	— ml
4.	—	—	— ml

Concordant volume = *x* ml (say).

CALCULATIONS

(i) Calculation of molarity of oxalic acid solution

From the overall balanced chemical equation, it is clear that 2 moles of $KMnO_4$ react with 5 moles of oxalic acid.

$$\therefore \quad \frac{M_{KMnO_4} \times V_{KMnO_4}}{M_{\text{oxalic acid}} \times V_{\text{oxalic acid}}} = \frac{2}{5}$$

$$\frac{0.01 \times x}{M_{\text{oxalic acid}} \times 20} = \frac{2}{5}$$

$$M_{\text{oxalic acid}} = \frac{x}{800}$$

(ii) Calculation of strength of oxalic acid solution (in g/l)

$$= \text{Molarity} \times \text{Molar mass}$$

$$= \frac{x}{800} \times 126 = y \text{ g/l (say).}$$

(iii) Calculation of percentage purity of oxalic acid

$$= \frac{\text{Strength of pure sample}}{\text{Strength of the given sample}} \times 100$$

$$= \frac{y}{4} \times 100$$

Instructions for the Preparation of Solutions

Provide the following :

1. Impure sample of oxalic acid
2. $KMnO_4$ solution (1.58 g/litre)
3. 2M H_2SO_4.

EXPERIMENT 11.9

AIM: ***The given solution has been prepared by dissolving 1.6 g of an alkali metal permanganate per litre of solution. Determine volumetrically the atomic mass of the alkali metal. Prepare 0.05 M Mohr's salt solution for titration.***

CHEMICAL EQUATIONS

Molecular equations

Let A represent the alkali metal and $AMnO_4$ represent alkali metal permanganate,

$$2AMnO_4 + 3H_2SO_4 \longrightarrow A_2SO_4 + 2MnSO_4 + 5[O]$$

$$FeSO_4.(NH_4)_2SO_4.6H_2O \longrightarrow FeSO_4 + (NH_4)_2SO_4 + 6H_2O$$

$$2FeSO_4 + H_2SO_4 + [O] \longrightarrow Fe_2(SO_4)_3 + H_2O$$

Ionic equations

$$MnO_4^- + 8H^+ + 5e^- \longrightarrow Mn^{2+} + 4H_2O$$

$$Fe^{2+} \longrightarrow Fe^{3+} + e^-] \times 5$$

$$MnO_4^- + 8H^+ + 5Fe^{2+} \longrightarrow 5Fe^{3+} + Mn^{2+} + 4H_2O$$

INDICATOR

$AMnO_4$ will act as a self-indicator.

END POINT

Colourless to permanent pink colour (*$AMnO_4$ in burette*).

PROCEDURE

1. Weigh exactly 4.90 g of Mohr's salt on a watch glass and dissolve in water to prepare exactly 250 ml of its 0.05 M solution using a 250 ml measured flask. Rinse the pipette with the 0.05 M Mohr's salt solution and pipette out 20.0 ml of it in a washed titration flask.
2. Rinse and fill the burette with the given $AMnO_4$ solution.
3. Add one test tube (~ 20 ml) full of dilute sulphuric acid (~ 2M) to the solution in titration flask.
4. Note the initial reading of the burette.
5. Now add $AMnO_4$ solution from the burette till a permanent light pink colour is just imparted to the solution in the titration flask.
6. Note the final reading of the burette.
7. Repeat the above steps 4–5 times to get three concordant readings.

OBSERVATIONS

Weight of watch glass = ... g

Weight of (watch glass + Mohr's salt) = ... g

Weight of Mohr's salt = 4.90 g

Volume of solution prepared = 250 ml

Molarity of Mohr's salt solution = 0.05 M

Volume of Mohr's salt solution taken for each titration = 20.0 ml

S. No.	Burette Readings		Volume of the $KMnO_4$ solution used
	Initial	Final	
1.	—	—	— ml
2.	—	—	— ml
3.	—	—	— ml
4.	—	—	— ml

Concordant volume = x ml (say).

CALCULATIONS

From the balanced chemical equation, it can be seen that 2 moles of $AMnO_4$ react with 10 moles of Mohr's salt.

$$\therefore \quad \frac{M_{AMnO_4} \times V_{AMnO_4}}{M_{Mohr's\ salt} \times V_{Mohr's\ salt}} = \frac{2}{10}$$

$$\frac{M_{AMnO_4} \times x}{0.05 \times 20} = \frac{2}{10}$$

Molarity of $AMnO_4$, $\quad M_{AMnO_4} = \frac{2}{10} \times \frac{1}{x} = \frac{2}{10x}$

Strength of $AMnO_4$ = 1.6 g/l.

Molecular mass of alkali metal permanganate

$$= \frac{\text{Strength (in g/l)}}{\text{Molarity}}$$

$$= \frac{1.6}{2/10x} = 8x \qquad ...(11.1)$$

But molecular mass of $AMnO_4$ = Atomic mass of A + Formula mass of MnO_4^-

$$= a + 119 \qquad ...(11.2)$$

From equations (11.1) and (11.2),

$$8x = a + 119$$

$$a = 8x - 119$$

Knowing the value of x, the atomic mass of A, can be calculated.

Instructions for the Preparation of Solutions

Provide the following :

1. $KMnO_4$ solution (1.58 g/litre)
   Label it as $AMnO_4$ solution
2. Mohr's salt
3. 2 M H_2SO_4.

EXPERIMENT 11.10

AIM: ***Determine the percentage composition of a mixture of sodium oxalate $\left(\begin{array}{l}COONa\\ |\\ COONa\end{array}\right)$ and oxalic acid $\left(\begin{array}{l}COOH\\ |\\ COOH\end{array}.2H_2O\right)$. Provided 0.01 M $KMnO_4$ solution.***

CHEMICAL EQUATIONS

Molecular Equations

$$2KMnO_4 + 3H_2SO_4 \longrightarrow K_2SO_4 + 2MnSO_4 + 3H_2O + 5[O]$$

$$\begin{array}{l}COOH\\ |\\ COOH\end{array}.2H_2O + [O] \longrightarrow 2CO_2 + 3H_2O$$

$$\begin{array}{l}COONa\\ |\\ COONa\end{array} + H_2SO_4 + [O] \longrightarrow 2CO_2 + H_2O + Na_2SO_4$$

Ionic Equations

$$MnO_4^- + 8H^+ + 5e^- \longrightarrow Mn^{2+} + 4H_2O\ ] \times 2$$

$$C_2O_4^{2-} \longrightarrow 2CO_2 + 2e^-\ ] \times 5$$

$$2MnO_4^- + 16H^+ + 5C_2O_4^{2-} \longrightarrow 2Mn^{2+} + 8H_2O + 10CO_2$$

INDICATOR

$KMnO_4$ is a self-indicator.

END POINT

Colourless to permanent pink (*$KMnO_4$ in burette*).

PROCEDURE

1. Weigh exactly 1.0 g of the given mixture of sodium oxalate and oxalic acid and dissolve in water to prepare exactly 250 ml of solution using a 250 ml measuring flask. Rinse the pipette with the given oxalate solution and pipette out 20.0 ml of it in a washed titration flask.
2. Rinse and fill the burette with the 0.01 M $KMnO_4$ solution.
3. Add one test tube (~ 20 ml) full of dilute sulphuric acid (~ 2 M) to the solution in titration flask.
4. Note the initial reading of the burette.
5. Heat the solution of titration flask to 60–70°C and run down $KMnO_4$ solution from the burette till a permanent light pink colour is imparted to the solution in the titration flask on addition of a last single drop of $KMnO_4$ solution.
6. Note the final reading of the burette.
7. Repeat the above steps 4–5 times to get three concordant reading.

OBSERVATIONS

Weight of watch glass = ... g

Weight of (watch glass + mixture) = ... g

Weight of mixture = 1.0 g

Volume of solution prepared = 250 ml

Molarity of $KMnO_4$ solution = 0.01 M

Volume of oxalate solution taken for each titration = 20.0 ml.

S. No.	Burette Readings		Volume of the $KMnO_4$ solution used
	Initial	Final	
1.	—	—	— ml
2.	—	—	— ml
3.	—	—	— ml
4.	—	—	— ml

Concordant volume = x ml (say).

CALCULATIONS

Strength of the prepared oxalate solution = 4.0 g/litre.

x ml of 0.01 M $KMnO_4$ react with 20.0 ml of the prepared oxalate solution.

From the ionic equation it is clear that 2 moles of MnO_4^- react with 5 moles of $C_2O_4^{2-}$ ions.

$$\frac{M_{MnO_4^-} \times V_{MnO_4^-}}{M_{C_2O_4^{2-}} \times V_{C_2O_4^{2-}}} = \frac{2}{5}$$

$$\frac{0.01 \times x}{M_{C_2O_4^{2-}} \times 20.0} = \frac{2}{5}$$

$$M_{C_2O_4^{2-}} = \frac{0.01 \times x \times 5}{20.0 \times 2} = \frac{x}{800}$$

$\therefore$ Total molarity of oxalate ions $= \dfrac{x}{800}$.

This molarity is due to oxalic acid as well as sodium oxalate.

Suppose strength of oxalic acid = a g/l

$\therefore$ Strength of sodium oxalate = $(4 - a)$ g/l

Molarity due to oxalic acid, $M_{\text{oxalic acid}} = \dfrac{a}{126}$

Molarity due to sodium oxalate, $M_{\text{sod. oxalate}} = \dfrac{4-a}{134}$

Total molarity of oxalate solution = $M_{\text{oxalic acid}} + M_{\text{sod. oxalate}}$

$$\frac{x}{800} = \frac{a}{126} + \frac{4-a}{134}$$

From this equation 'a' can be calculated. Knowing 'a', the percentage composition of the mixture can be calculated.

% of oxalic acid $= \dfrac{a}{4} \times 100$

% of sodium oxalate $= \dfrac{4-a}{4} \times 100$

Instructions for the Preparation of Solutions

Provide the following solutions :

1. $KMnO_4$ solution (1.58 g/litre)
2. A mixture of oxalic acid and sodium oxalate crystals
3. 2 M H_2SO_4.

EXPERIMENT 11.11

AIM: ***You are provided with a partially oxidised sample of ferrous sulphate ($FeSO_4 . 7H_2O$) crystals. Prepare a solution by dissolving 14.0 g of these crystals per litre and determine the percentage oxidation of the given sample. Given 0.01 M $KMnO_4$ solution.***

CHEMICAL EQUATIONS

Molecular Equation

$$2KMnO_4 + 3H_2SO_4 \longrightarrow K_2SO_4 + 2MnSO_4 + 3H_2O + 5[O]$$
$$2FeSO_4 + H_2SO_4 + [O] \longrightarrow Fe_2(SO_4)_3 + H_2O\,] \times 5$$
$$\overline{2KMnO_4 + 10FeSO_4 + 8H_2SO_4 \longrightarrow K_2SO_4 + 2MnSO_4 + 8H_2O + Fe_2(SO_4)_3}$$

Ionic equation

$$MnO_4^- + 8H^+ + 5e^- \longrightarrow Mn^{2+} + 4H_2O$$
$$Fe^{2+} \longrightarrow Fe^{3+} + e^-\,] \times 5$$
$$\overline{MnO_4^- + 8H^+ + 5Fe^{2+} \longrightarrow 5Fe^{3+} + Mn^{2+} + 4H_2O}$$

THEORY

Since the given sample contains partially oxidized ferrous sulphate, it contains both ferrous ions, Fe^{2+} (unoxidised) and ferric ions Fe^{3+} (oxidised). The strength of partially oxidised sample is known. The solution of partially oxidised $FeSO_4$ of known strength is titrated against standard $KMnO_4$ solution to determine the molarity and strength of the unoxidised ferrous sulphate. From this the percentage oxidation of the sample can be calculated.

INDICATOR

$KMnO_4$ is a self-indicator.

END POINT

Colourless to permanent pink colour (*$KMnO_4$ in burette*).

PROCEDURE

1. Weigh exactly 3.50 g of the given sample of ferrous sulphate on a watch glass and dissolve in water to prepare exactly 250 ml of solution using a 250 ml measuring flask. Rinse and fill the pipette with prepared ferrous sulphate solution and pipette out 20.0 ml of it in a washed titration flask.
2. Rinse and fill the burette with the 0.01 M $KMnO_4$ solution.
3. Add one test-tube (~ 20 ml) full of dilute sulphuric acid (~ 2 M) to the solution in the titration flask.
4. Note the initial reading of the burette.
5. Now add $KMnO_4$ solution from the burette till a permanent light pink colour is imparted to the solution in the titration flask on addition of a last single drop of $KMnO_4$ solution.
6. Note the final reading of the burette.
7. Repeat the above steps 4–5 times to get three concordant readings.

OBSERVATIONS

Weight of watch glass = ... g

Weight of (watch glass + ferrous sulphate) = ... g

Weight of ferrous sulphate crystals = 3.50 g

Volume of solution prepared = 250 ml

Strength of solution = 14.0 g/litre

Molarity of $KMnO_4$ solution = 0.01 M.

Volume of ferrous sulphate solution taken for each titration = 20.0 ml.

S. No.	Burette Readings		Volume of the $KMnO_4$ solution used
	Initial	Final	
1.	—	—	— ml
2.	—	—	— ml
3.	—	—	— ml
4.	—	—	— ml

Concordant volume = x ml (say).

CALCULATIONS

Molarity of the standard $KMnO_4$ solution = 0.01 M

Volume of 0.01 M $KMnO_4$ solution required for the oxidation of 20.0 ml of the given ferrous sulphate solution = x ml.

From the chemical equations, it is clear that 2 moles of $KMnO_4$ react with 10 moles of ferrous sulphate.

$$\therefore \quad \frac{M_{KMnO_4} \times V_{KMnO_4}}{M_{FeSO_4} \times V_{FeSO_4}} = \frac{2}{10}$$

$$\frac{0.01 \times x}{M_{FeSO_4} \times 20.0} = \frac{2}{10}$$

$$M_{FeSO_4} = \frac{0.01 \times x \times 10}{20.0 \times 2} = \frac{x}{400}$$

$\therefore$ Molarity of unoxidized ferrous sulphate = $\frac{x}{400}$

Strength of unoxidized ferrous sulphate (in g/l)

$$= \text{Molarity} \times \text{Molar mass of } FeSO_4 .7H_2O$$

$$= \frac{x}{400} \times 278 = y \text{ g/l}$$

Total strength of partially oxidised sample = 14 g/l

∴ Strength of oxidised ferrous sulphate = (14 – y) g/l

$$\% \text{ Oxidation} = \frac{14 - y}{14} \times 100.$$

Instructions for the Preparation of Solutions

Provide the following :

1. $KMnO_4$ solution (1.58 g/litre)
2. $FeSO_4.7H_2O$
3. 2 M H_2SO_4.

EXPERIMENT 11.12

AIM: ***Calculate the percentage of Fe^{2+} ions in a sample of ferrous sulphate. Prepare a solution of the given sample having strength exactly equal to 14.0 g/litre. Provided 0.01 M $KMnO_4$.***

CHEMICAL EQUATIONS

Same as in Expt. 11.11.

Indicator, End point and **Procedure** are also same as in Expt. 11.11.

OBSERVATIONS

Weight of watch glass = ... g

Weight of (watch glass + ferrous sulphate) = ... g

Weight of ferrous sulphate = 3.50 g

Volume of solution prepared = 250 ml

Molarity of $KMnO_4$ = 0.01 M

Volume of ferrous sulphate solution taken for each titration = 20.0 ml.

S. No.	*Burette Readings*		*Volume of the $KMnO_4$ solution used*
	Initial	*Final*	
1.	—	—	— ml
2.	—	—	— ml
3.	—	—	— ml
4.	—	—	— ml

Concordant volume = x ml (say).

CALCULATIONS

Volume of 0.01 M $KMnO_4$ solution required for the oxidation of 20.0 ml of the prepared ferrous sulphate solution = x ml.

From the equations it is clear that 2 moles of $KMnO_4$ react with 10 moles of ferrous sulphate.

$$\therefore \quad \frac{M_{KMnO_4} \times V_{KMnO_4}}{M_{FeSO_4} \times V_{FeSO_4}} = \frac{2}{10}$$

$$\frac{0.01 \times x}{M_{FeSO_4} \times 20.2} = \frac{2}{10}$$

$$M_{FeSO_4} = \frac{0.01 \times x \times 10}{20.0 \times 2} = \frac{x}{400}$$

Molarity of Fe^{2+} ions $= \text{Molarity of ferrous sulphate} = \frac{x}{400}$

Strength of Fe^{2+} ions $= \text{Molarity} \times \text{Formula mass}$

$$= \frac{x}{400} \times 56 \text{ g/litre}$$

$$= y \text{ g/litre (say)}$$

Percentage of Fe^{2+} ions in the given sample of ferrous sulphate

$$= \frac{\text{Strength of } Fe^{2+} \text{ ions in g/litre}}{\text{Strength of ferrous sulphate in g/litre}} \times 100$$

$$= \frac{y}{14} \times 100.$$

Instructions for the Preparation of Solutions

Provide the following :

1. $KMnO_4$ solution (1.58 g/litre)
2. $FeSO_4.7H_2O$ crystals
3. 2 M H_2SO_4.

EXERCISES

Q. 1. *Prepare a standard solution of M/50 $FeSO_4(NH_4)_2SO_4.6H_2O$ (Mohr's salt). Using this solution find out the molarity of the given solution of $KMnO_4$.*

Q. 2. *Prepare M/50 solution of oxalic acid. Using this solution find out the molarity and strength of the given solution of $KMnO_4$.*

Q. 3. *Prepare a solution of ferrous ammonium sulphate containing exactly 4.9 g of the salt per 250 ml of solution. Using this solution determine the concentration of $KMnO_4$ in g/litre in the given solution.*

Q. 4. *Prepare M/20 solution of oxalic acid. Using this solution find out percentage purity of impure sample of $KMnO_4$, 3.5 g of which have been dissolved per litre.*

Q. 5. *Prepare M/50 ferrous ammonium sulphate solution. With its help, find out the percentage purity of impure sample of $KMnO_4$, 3.6 g of which have been dissolved per litre.*

Q. 6. *Prepare M/20 oxalic acid solution. You are provided two solutions of $KMnO_4$, A and B. Find out volumetrically which solution, (A or B) is more concentrated. Report the strength of more concentrated solution in g/litre.*

Q. 7. *You are provided with a solution of alkali metal permanganate, $AMnO_4$ containing 3.15 g of it per litre of the solution. Prepare M/20 oxalic acid solution and using this solution determine the atomic mass of the alkali metal 'A'.*

Q. 8. *Prepare a solution containing exactly 3.0 g of* $\begin{array}{l}COOH\\ | \\ COOH\end{array} \cdot nH_2O$ *per 500 ml of solution. Find out volumetrically the value of n. Provided M/50 $KMnO_4$ solution.*

Q. 9. *Determine volumetrically the percentage purity of a given sample of sodium oxalate. Provided M/50 $KMnO_4$ solution.*

VIVA QUESTIONS WITH ANSWERS

Q. 1. What is a standard solution ?

Ans. A solution whose strength is known is called a standard solution.

Q. 2. What is a normal solution ?

Ans. A solution containing one gram-equivalent mass of the solute per litre of the solution is called a normal solution.

Q. 3. What is the equivalent mass of $KMnO_4$ when it acts as oxidizing agent in acidic medium ?

Ans. $KMnO_4$ loses 5 electrons per molecule, when it acts as oxidizing agent in the presence of acids. Therefore, its equivalent mass is one-fifth of its molecular mass.

$$\text{Equivalent mass} = \frac{\text{Molecular mass}}{5} = \frac{158}{5} = \mathbf{31.6.}$$

Q. 4. Is sodium hydroxide a primary standard ?

Ans. No.

Q. 5. Are 'molality' and 'molarity' the same ?

Ans. No, molality of a solution is defined as the number of moles of solute present in one kg of the solvent whereas molarity tells us about the number of moles of the solute present per litre of the solution.

Q. 6. What would be the normality of 0.10 M $KMnO_4$?

Ans. It will be $0.1 \times 5 = 0.5$ N.

Q. 7. What volume of 10 M HCl must be diluted with water to get 1 L of 1 M HCl ?

Ans. 0.1 L.

Q. 8. What is the basicity of H_2SO_4 ?

Ans. 2.

Q. 9. What is the relationship between normality (N), molarity (M), molecular mass and equivalent mass ?

Ans. Normality × Equivalent Mass = Molarity × Molecular Mass.

Q. 10. Why front door of the balance is closed during weighing ?

Ans. Opening the front door causes vibrations in the pan due to operator's breath which leads to inaccurate results.

Q. 11. What is the maximum weight that can be weighed in a chemical balance ?

Ans. 100 grams.

Q. 12. What is the weight of a rider ?

Ans. 10 mg.

Q. 13. What is the use of a rider ?

Ans. A rider is used for weights less than 10 mg.

Q. 14. What is the principle of volumetric analysis ?

Ans. In volumetric analysis, the concentration of a solution is determined by allowing a known volume of the solution to react, quantitatively with another solution of known concentration.

Q. 15. What is titration ?

Ans. The process of adding one solution from the burette to another in the conical flask in order to complete the chemical reaction involved, is known as titration.

Q. 16. What is indicator ?

Ans. Indicator is a chemical substance which changes colour at the end point.

Q. 17. What is end point ?

Ans. The stage during titration at which the reaction is just complete is known as the end point of titration.

Q. 18. Why a titration flask should not be rinsed ?

Ans. This is because during rinsing some liquid will remain sticking to the titration flask therefore the pipetted volume taken in the titration flask will increase.

Q. 19. What are primary and secondary standard substances ?

Ans. A substance is known as primary standard if it is available in high degree of purity, if it is stable and unaffected by air, if it does not gain or lose moisture in air, if it is readily soluble and its solution in water remains as such for long time.

On the other hand, a substance which does not possess the above characteristics is called a secondary standard substance. Primary standards are crystalline oxalic acid, anhydrous Na_2CO_3, Mohr's salt, etc.

Q. 20. Burette and pipette must be rinsed with the solution with which they are filled, why ?

Ans. The burette and pipette are rinsed with the solution with which they are filled in order to remove any water sticking to their sides, which otherwise would decrease the conc. of the solutions to be taken in them.

Q. 21. It is customary to read lower meniscus in case of colourless and transparent solutions and upper meniscus in case of highly coloured solutions, why ?

Ans. Because it is easy to read the lower meniscus in case of colourless solutions, while the upper meniscus in case of coloured solutions. In case of coloured solutions lower meniscus is not visible clearly.

Q. 22. What is a molar solution ?

Ans. A molar solution is a solution, a litre of which contains one gm-mole of the substance. This is symbolised as 1 M.

Q. 23. Why the last drop of solution must not be blown out of a pipette ?

Ans. Since the drops left in the jet end is extra of the volume measured by the pipette.

Q. 24. Pipette should never be held from its bulb, why ?

Ans. The body temperature may expand the glass and introduce an error in the measurement volume.

Q. 25. What is acidimetry and alkalimetry ?

Ans. It is the branch of volumetric analysis involving chemical reaction between an acid and a base.

Q. 26. What is permanganometry ?

Ans. Redox titrations involving $KMnO_4$ as the oxidising agent are called permanganometric titrations.

Q. 27. Which is an oxidising agent and a reducing agent in the reaction between $KMnO_4$ and $FeSO_4$?

Ans. $KMnO_4$ acts as oxidising agent and $FeSO_4$ acts as reducing agent.

Q. 28. What is the indicator used in $KMnO_4$ titration ?

Ans. No external indicator is used because $KMnO_4$ acts as a self-indicator.

Q. 29. Why does $KMnO_4$ act itself as an indicator ?

Ans. In the presence of dilute sulphuric acid, $KMnO_4$ reacts with reducing agent (oxalic acid or ferrous sulphate). When all the reducing agent has been oxidised, the excess of $KMnO_4$ is not decomposed and imparts pink colour to the solution.

Q. 30. What is the end point in $KMnO_4$ titrations ?

Ans. From colourless to permanent light pink.

Q. 31. Why is Mohr's salt preferred as a primary standard over ferrous sulphate in volumetric analysis ?

Ans. This is because of the fact that Mohr's salt is stable and is not readily oxidised by air. Ferrous sulphate gets oxidised to ferric sulphate.

Q. 32. Why are a few drops of dilute sulphuric acid added while preparing a standard solution of Mohr's salt ?

Ans. Few drops of H_2SO_4 are added to prevent the hydrolysis of ferrous sulphate.

Q. 33. Why a burette with rubber pinch cock should not be used in $KMnO_4$ titrations ?

Ans. Because $KMnO_4$ attacks rubber.

Q. 34. Sometimes a brown ppt. is observed in $KMnO_4$ titrations. Why ?

Ans. It is due to insufficient quantity of dil. sulphuric acid. Brown coloured ppt. ($MnO_2.H_2O$) is formed due to the incomplete oxidation of $KMnO_4$.

$$2KMnO_4 + H_2O \longrightarrow 2KOH + \underset{\text{Brown ppt.}}{2MnO_2} + 3[O]$$

Q. 35. Why should you heat the oxalic acid solution to about 60–70°C before titrating with $KMnO_4$ solution ?

Ans. In cold, the reaction is very slow due to the slow formation of Mn^{2+} ions. Oxalic acid is heated to speed up the liberation of Mn^{2+} ions which then autocatalyses the reaction and thus the reaction proceeds rapidly. This also serves the purpose of expelling the carbondioxide evolved during the reaction which otherwise does not allow the reaction to go to completion.

CHAPTER 12

Qualitative Analysis

Analytical chemistry deals with qualitative and quantitative analysis of the substances. In qualitative analysis, the given substance is analyzed for the nature of the substance and its constituents. Qualitative analysis of inorganic salts involves identification of cations and anions present in the salt or a mixture of salts. Physical procedures like noting the colour, smell or taste of the substance have very limited scope because of the corrosive, poisonous nature of the chemical compounds. Therefore, what one has to resort to is the chemical analysis of the substance that has to be carried out along with the physical examination of the compound under consideration.

The common procedure for testing any unknown sample is to make its solution and then test this solution for the ions present in it. There are separate procedures for detecting cations and anions. Therefore, qualitative analysis is studied under cation analysis and anion analysis. The systematic procedure for qualitative analysis of an inorganic salt involves the following steps:

(*a*) Preliminary Tests

1. Physical appearance (colour and smell).
2. Dry heating test.
3. Charcoal cavity test.
4. Charcoal cavity and cobalt nitrate test.
5. Flame test.
6. Borax bead test.
7. Dilute acid test.
8. Potassium permanganate test.
9. Concentrated sulphuric acid test.
10. Tests for sulphate, phosphate and borate.

(*b*) Wet Tests for Acid Radical.

(*c*) Wet Tests (Group Analysis) for Basic Radical.

12.01. PHYSICAL EXAMINATION OF THE SALT

The physical examination of the unknown salt involves the study of *colour, smell and density*. The test is not much reliable, but is certainly helpful in identifying some coloured cations. Characteristic smell helps to identify some ions such as ammonium, acetate and sulphide.

(See Table 12.1 on next page)

Notes:

1. *If you have touched any salt, wash your hands at once. It may be corrosive to skin.*
2. *Never taste any salt, it may be poisonous. Salts of arsenic and mercury are highly poisonous.*
3. *Salts like sodium sulphide, sodium nitrite, potassium nitrite, develop a yellow colour.*

Table 12.1. Physical Examination

Experiment	*Observations*	*Inference*
1. *Colour*	Blue or Bluish green	Cu^{2+}
	Greenish	Ni^{2+}
	Light green	Fe^{2+}
	Dark brown	Fe^{3+}
	Pink	Co^{2+}
	Light pink, flesh colour or earthy colour	Mn^{2+}
	White	Shows the absence of Cu^{2+}, Co^{2+}, Ni^{2+}, Fe^{2+}, Fe^{3+}, Mn^{2+}
2. *Smell* Take a pinch of the salt between your fingers and rub with a drop of water	Ammoniacal smell	NH_4^+
	Vinegar like smell	CH_3COO^-
	Smell like that of rotten eggs	S^{2-}
3. *Density*	(*i*) Heavy	Salt of Pb^{2+}, or Ba^{2+}
	(*ii*) Light fluffy powder	Carbonate
4. *Deliquescence*	Salt absorbs moisture and becomes paste like	(*i*) If coloured, may be $Cu(NO_3)_2$, $FeCl_3$ (*ii*) If colourless, may be $Zn(NO_3)_2$, chlorides of Zn^{2+}, Mg^{2+} etc.

12.02. DRY HEATING TEST

This test is performed by heating a small amount of salt in a *dry test tube*. Quite valuable information can be gathered by carefully performing and noting the observations here. On heating, some salts undergo decomposition, thus, evolving the gases or may undergo characteristic changes in the colour of residue. These observations are tabulated in Table 12.2 along with the inferences that you can draw.

Table 12.2. Dry Heating Test

Observations	*Inference*
1. Gas evolved	
(*a*) Colourless and odourless gas	
CO_2 gas—turns lime water milky	CO_3^{2-} or $C_2O_4^{2-}$ present
(*b*) Colourless gas with odour	
(*i*) *H_2S gas*—Smells like rotten eggs, turns lead acetate paper black.	Hydrated S^{2-}

Observations	Inference
(ii) SO_2 gas—characteristics suffocating smell, turns acidified potassium dichromate paper (or solution) green.	SO_3^{2-}
(iii) HCl gas—Pungent smell, white fumes with ammonia, white ppt with silver nitrate solution.	Cl^-
(iv) Acetic acid vapours—Characteristic vinegar like smell.	CH_3COO^-
(v) NH_3 gas—Characteristic smell, gives white fumes when a glass rod dipped in dilute HCl is brought near the mouth of the test tube, turns Nessler's solution brown.	NH_4^+
(c) Coloured gases—Pungent smell	
(i) NO_2 gas—Reddish brown, turns ferrous sulphate solution black.	NO_2^- or NO_3^-
(ii) Cl_2 gas—Greenish yellow, turns starch-iodide paper blue.	Cl^-
(iii) Br_2 vapours—Reddish brown, turns starch paper orange yellow.	Br^-
(iv) I_2 vapours—Dark violet, turns starch paper blue.	I^-
2. Sublimate formed	
(a) White sublimate	NH_4^+
(a) Black sublimate accompanied by violet vapours	I^-
3. Decrepitation	
The salt decrepitates. (makes crackling sound)	A salt having no water of crystallisation may be present For example, $Pb(NO_3)_2$, NaCl, KBr.
4. Swelling	
The salt swells up into voluminous mass.	PO_4^{3-} indicated
5. Residue	
(i) Yellow when hot white when cold	Zn^{2+} indicated
(ii) Brown when hot and yellow when cold	Pb^{2+} indicated
(iii) White salt becomes black on heating	CH_3COO^- indicated
(iv) White residue, glows on heating	Ba^{2+}, Sr^{2+}, Ca^{2+}, Mg^{2+}, Al^{3+}, etc. may be present.
(v) Original salt blue becomes white on heating	Hydrated $CuSO_4$ indicated
(vi) Coloured salt becomes brown or black on heating.	Co^{2+}, Cu^{2+}, Mn^{2+} indicated.

Notes:

1. *Use a perfectly dry test-tube for performing this test. While drying a test-tube, keep it in slanting position with its mouth slightly downwards so that the drops of water which condense on the upper cooler parts, do not fall back on the hot bottom, as this may break the tube.*
2. *For testing a gas, a filter paper strip dipped in the appropriate reagent is brought near the mouth of the test tube or alternatively the reagent is taken in a gas-detector and the gas is passed through it [Fig. 12.1].*

3. *Do not heat the tube strongly at one point as it may break.*

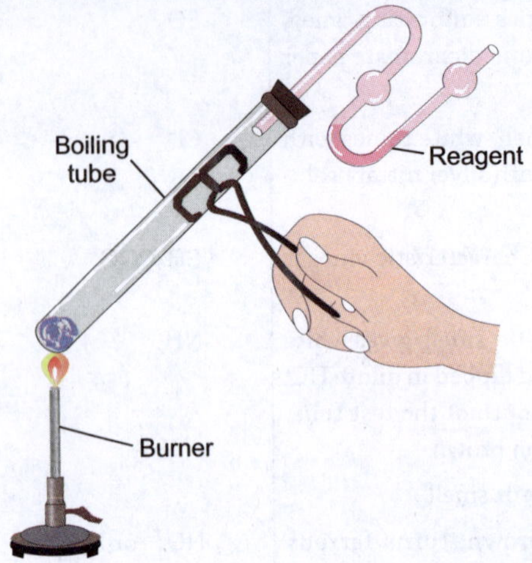

Fig. 12.1. Testing a gas.

12.03. CHARCOAL CAVITY TEST

This test is based on the fact that metallic carbonates when heated in a charcoal cavity decompose to give corresponding oxides. The oxides appear as coloured incrustation or residue in the cavity. In certain cases, the oxides formed partially undergo reduction to the metallic state producing metallic beads or scales.

Examples:

(*a*) $ZnSO_4 + Na_2CO_3 \longrightarrow ZnCO_3 + Na_2SO_4$

$ZnCO_3 \longrightarrow \underset{\text{Yellow when hot, white when cold}}{ZnO} + CO_2\uparrow$

(*b*) $Pb(NO_3)_2 + Na_2CO_3 \longrightarrow PbCO_3 + 2NaNO_3$

$PbCO_3 \longrightarrow PbO + CO_2\uparrow$

$PbO + C \longrightarrow \underset{\text{Bead}}{Pb} + CO\uparrow$

(*c*) $CuSO_4 + Na_2CO_3 \longrightarrow CuCO_3 + Na_2SO_4$

$CuCO_3 \longrightarrow CuO + CO_2\uparrow$

$CuO + C \longrightarrow \underset{\text{Reddish scales}}{Cu} + CO\uparrow$

PROCEDURE

While performing charcoal cavity test, make a small cavity on a charcoal block with the help of borer as shown in Fig. 12.2. Mix small amount of salt with double its quantity of sodium carbonate. Place it in the cavity made on the block of charcoal. Moisten with a drop of water and direct the reducing flame of the bunsen burner on the cavity by means of a mouth blowpipe as shown in Fig. 12.3. Heat strongly for sometime and draw inference according to the Table 12.3.

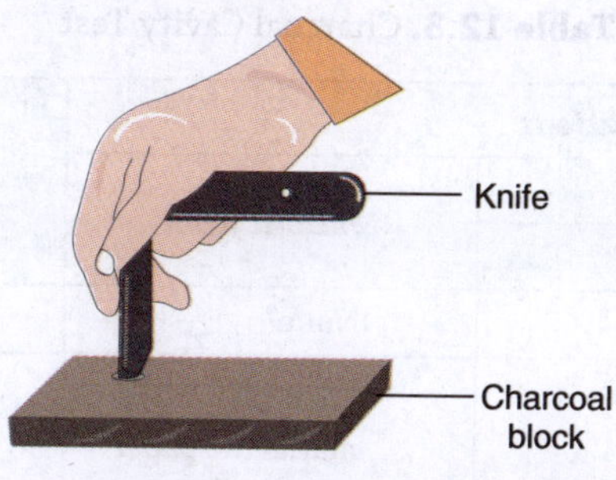

Fig. 12.2. Making bore on a charcoal block.

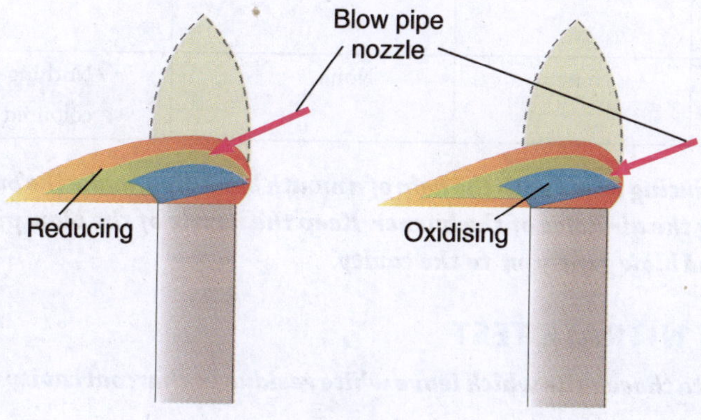

Fig. 12.3. Directing flame with blow pipe.

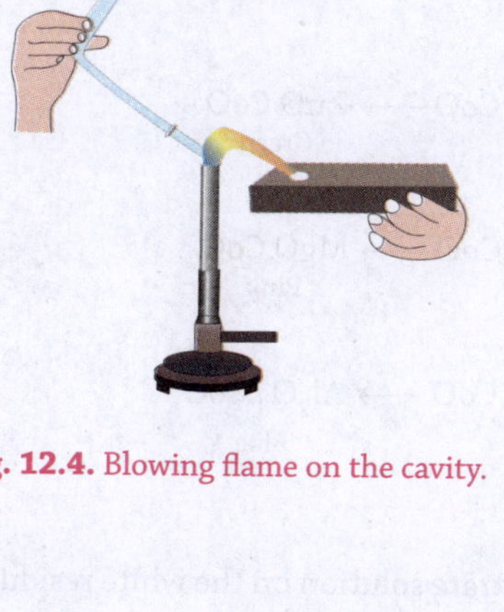

Fig. 12.4. Blowing flame on the cavity.

Table 12.3. Charcoal Cavity Test

Observations			***Inference***
Incrustation or Residue		***Metallic bead***	
Hot	***Cold***		
Yellow	White	None	Zn^{2+}
Brown	Yellow	Grey bead which marks the paper	Pb^{2+}
None	None	Red beads or scales	Cu^{2+}
White residue which glows	None	None	Ba^{2+}, Ca^{2+}, Mg^{2+}
Black	None	None	Nothing definite—generally coloured salt

To obtain a reducing flame with the help of a mouth blow pipe, make the bunsen burner flame luminous by closing the air holes of the burner. Keep the nozzle of the blow pipe just outside the flame (Fig. 12.4) and blow gently on to the cavity.

12.04. COBALT NITRATE TEST

This test is applied to those salts which leave white residue in charcoal cavity test.

The test is based on the fact that cobalt nitrate decomposes on heating to give cobalt oxide, CoO. This combines with the metallic-oxides, present as white residue in the charcoal cavity forming coloured compounds. For example, when a magnesium salt undergoes charcoal cavity test, a white residue of MgO is left behind. This on treatment with cobalt nitrate and subsequent heating forms a double salt of the formula MgO.CoO which is pink in colour. In addition to metallic oxides, phosphates and borates also react with cobalt oxide to form $Co_3(PO_4)_2$ and $Co_3(BO_3)_2$ which are blue in colour.

Some of the reactions involved are given below:

$$2Co(NO_3)_2 \xrightarrow{\Delta} 2CoO + 4NO_2 + O_2$$

(i) Zinc salt:

$$ZnO + CoO \longrightarrow \underset{\text{Green}}{ZnO.CoO}$$

(ii) Magnesium salt:

$$MgO + CoO \longrightarrow \underset{\text{Pink}}{MgO.CoO}$$

(iii) Aluminium salt:

$$Al_2O_3 + CoO \longrightarrow \underset{\text{Blue}}{Al_2O_3.CoO}$$

PROCEDURE

Put one or two drops of cobalt nitrate solution on the white residue left after charcoal cavity test. Heat for one or two minutes by means of a blow pipe in oxidising flame. Observe the colour of the residue and draw inferences from Table 12.4.

Table 12.4. Cobalt Nitrate-Charcoal Cavity Test

Colour of the residue	*Inference*
Green	Zn^{2+}
Pink	Mg^{2+}
Blue	Al^{3+} or PO_4^{3-}
Black	It is due to the formation of CoO. No definite indication.

Notes:

1. *In order to obtain oxidising flame, keep the air holes of the burners open and place nozzle of the blowpipe about one third within the flame.*
2. *Perform this test only if the residue in the charcoal cavity test is white.*
3. *Do not put more than 2 drops of cobalt nitrate on the white residue. Excess cobalt nitrate may decompose to give cobalt oxide which is black in colour.*
4. *Use dilute solution of cobalt nitrate.*

12.05. FLAME TEST

Certain salts on reacting with conc. hydrochloric acid from their chlorides, that are volatile in non-luminous flame. Their vapours impart characteristic colour to the flame. This colour can give reliable information of the presence of certain basic radicals.

For proceeding to this test, the paste of the mixture with conc. hydrochloric acid is introduced into the flame with the help of platinum wire (Fig. 12.5).

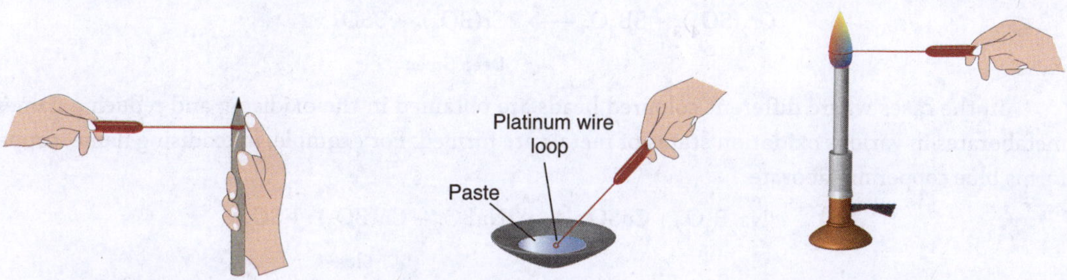

(a) Making a loop of platinum wire. (b) Dipping the platinum wire in the paste of salt and HCl. (c) Introducing the wire in the flame.

Fig. 12.5. Flame test.

PROCEDURE

Clean the platinum wire by dipping it in some conc. HCl taken on a watch glass and then heating strongly in the flame. This process is repeated till the wire imparts no colour to the flame. Now prepare a paste of the mixture with conc. HCl on a clean watch glass. Place small amount of this paste on platinum wire loop and introduce it into the flame. Note the colour imparted to the flame.

Table 12.5. Flame Test

Colour of the flame	Inference
1. Brick-red (not persistent)	Ca^{2+}
2. Crimson-red (persistent)	Sr^{2+}
3. Persistent grassy-green (appears after prolonged heating)	Ba^{2+}
4. Bright-bluish green	Cu^{2+}
5. Green flashes	Zn^{2+} or Mn^{2+}
6. Dull bluish-white	Pb^{2+}

12.06. BORAX BEAD TEST

This test is performed only for coloured salts.

Borax, $Na_2B_4O_7.10H_2O$, on heating gets fused and loses water of crystallisation. It swells up into a fluffy white porous mass which then melts into a colourless liquid which later forms a clear transparent glassy bead consisting of boric anhydride and sodium metaborate.

$$Na_2B_4O_7.10H_2O \longrightarrow Na_2B_4O_7 + 10H_2O\uparrow$$

$$Na_2B_4O_7 \longrightarrow \underset{\text{Boric anhydride}}{B_2O_3} + \underset{\text{Sodium metaborate}}{2NaBO_2}$$

Boric anhydride is non-volatile. When it is reacted with coloured metallic salt, a characteristic coloured bead of metal metaborate is formed.

$$Cr_2(SO_4)_3 + 3B_2O_3 \longrightarrow \underset{\text{Deep Green}}{2Cr(BO_2)_3} + 3SO_3$$

In the cases where different coloured beads are obtained in the oxidising and reducing flames, metaborates in various oxidation states of metals are formed. For example, in oxidising flame, copper forms blue copper metaborate.

$$Na_2B_4O_7 + CuSO_4 \longrightarrow 2NaBO_2 + \underset{\text{Blue}}{Cu(BO_2)_2} + SO_3$$

In reducing flame cupric metaborate is reduced to metallic copper, which is red and opaque.

$$2Cu(BO_2)_2 + 4NaBO_2 + 2C \longrightarrow \underset{\text{Red opaque}}{2Cu} + 2Na_2B_4O_7 + 2CO.$$

PROCEDURE

Borax, $Na_2B_4O_7.10H_2O$ is heated in the loop of platinum wire, it swells and forms transparent colourless glassy bead. When this hot bead is touched with small amount of coloured salt and is heated again, it acquires a characteristic colour. The colour of bead gives indication of the type of the cation present. The colour of the bead is noted separately in oxidising and in reducing flame (Fig. 12.6).

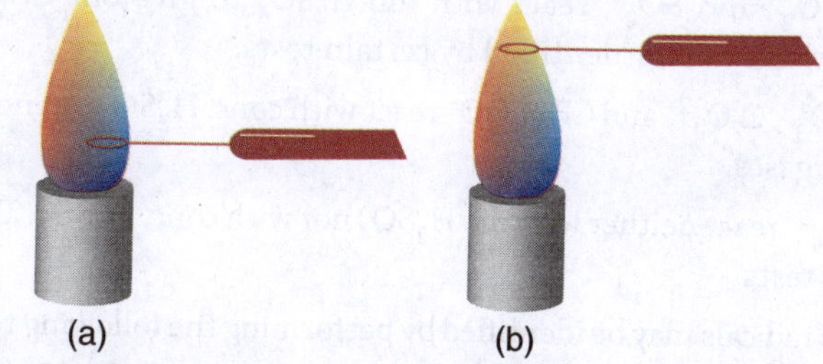

Fig. 12.6. Borax bead test *(a)* Heating in reducing flame, *(b)* Heating in oxidising flame.

Table 12.6. Borax Bead Test

Colour of the bead		Inference
In Oxidising flame	***In Reducing flame***	
1. Green when hot, light blue when cold.	Colourless when hot, opaque red when cold.	Cu^{2+}
2. Yellowish brown when hot, pale yellow when cold.	Green, hot and cold.	Fe^{2+} or Fe^{3+}
3. Amethyst (pinkish violet) in both hot and cold.	Colourless, hot and cold.	Mn^{2+}
4. Violet when hot, pale brown when cold.	Grey or black when hot and opaque when cold.	Ni^{2+}
5. Deep blue in both hot and cold.	Deep blue in both hot and cold.	Co^{2+}

To remove the bead from platinum wire, heat the bead to redness. Tap the rod with finger stroke, till the bead jumps off (Fig. 12.7).

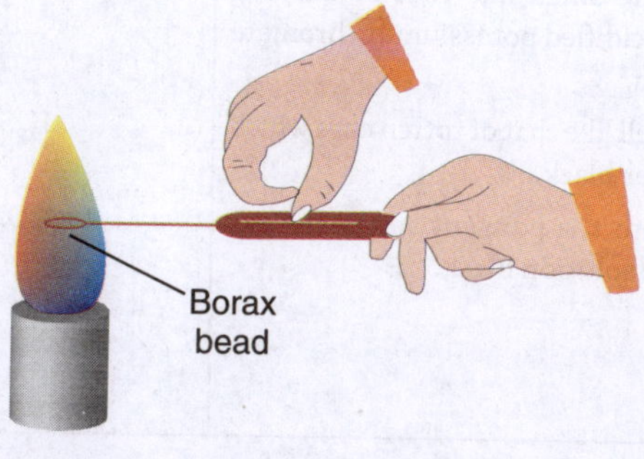

Fig. 12.7. Removing bead from platinum wire.

IDENTIFICATION OF ACID RADICALS (ANIONS)

The identification of the acid radicals is first done on the basis of preliminary tests. Dry heating test is one of the preliminary tests performed earlier which may give some important information about the acid radical present. The other preliminary tests are based upon the fact that:

1. CO_3^{2-}, S^{2-}, NO_2^- and SO_3^{2-} react with dil. H_2SO_4 to give out CO_2, H_2S, NO_2 and SO_2 gases respectively which can be identified by certain tests.
2. Cl^-, Br^-, I^-, NO_3^-, $C_2O_4^{2-}$ and CH_3COO^- react with conc. H_2SO_4 but not with dil. H_2SO_4 to produce characteristic gases.
3. SO_4^{2-} and PO_4^{3-} react neither with dil. H_2SO_4 nor with conc. H_2SO_4. These are therefore, identified by individual tests.

Thus, the acid radicals may be identified by performing the following tests in the order given below:

(i) *Dil. H_2SO_4 test.* Treat a pinch of the salt with dil. H_2SO_4 and identify the gas evolved.

(ii) *Conc. H_2SO_4 test.* If no action takes place with dil. H_2SO_4, warm a pinch of the salt with conc. H_2SO_4 and identify the gas evolved.

(iii) *Independent Group (SO_4^{2-} and PO_4^{3-}).* If the salt does not react with dil. H_2SO_4 as well as with conc. H_2SO_4, test for SO_4^{2-} and PO_4^{3-} by performing their individual tests.

Let us now discuss these tests in detail one by one.

12.07. DILUTE SULPHURIC ACID TEST

Take a small quantity of the salt in a test tube and add 1–2 ml of dilute sulphuric acid. Identify the gas and draw inferences from Table 12.7.

Table 12.7. Dilute Sulphuric Acid Test

Observations	*Inference*	
	Gas evolved	*Radical*
1. Colourless, odourless gas with *brisk effervescence,* which turns lime water milky.	CO_2	CO_3^{2-}
2. Colourless gas with the smell like that of burning sulphur which turns acidified potassium dichromate paper or solution green.	SO_2	SO_3^{2-}
3. Colourless gas with smell like that of rotten eggs which turns lead acetate paper black.	H_2S	S^{2-}
4. Reddish brown gas with the pungent smell. The gas turns ferrous sulphate solution black.	NO_2	NO_2^-
5. No gas evolved.	—	CO_3^{2-}, SO_3^{2-}, S^{2-} NO_2^- absent

Notes:

1. *Do not treat the salt with a large quantity of dilute acid.*
2. *Do not heat the salt with dilute acid.*
3. *Some acetates may react with dilute sulphuric acid and produce vapours of acetic acid having vinegar-like smell.*

Chemical Reactions Involved in Dil. H_2SO_4 Test

Dilute H_2SO_4 (or dilute HCl) decomposes carbonates, sulphides and nitrites in cold to give gases. These gases on identification indicate the nature of the acid radical present in the salt.

1. ***Carbonates.*** On treating the solid carbonate, carbon dioxide gas is given off in the cold with brisk effervescence, which turns lime water milky.

$$CaCO_3 + H_2SO_4 \longrightarrow CaSO_4 + H_2O + CO_2\uparrow$$

$$Ca(OH)_2 + CO_2 \longrightarrow \underset{\text{Milkiness}}{CaCO_3\downarrow} + H_2O$$

2. ***Sulphides.*** Sulphides when treated with dil. H_2SO_4 give H_2S gas, which turns lead acetate paper black due to the formation of lead sulphide.

$$ZnS + H_2SO_4 \longrightarrow ZnSO_4 + H_2S\uparrow$$

3. ***Sulphites.*** On treating solid sulphite with dil. H_2SO_4, SO_2 gas is evolved, which turns potassium dichromate solution green.

$$Na_2SO_3 + H_2SO_4 \longrightarrow Na_2SO_4 + SO_2\uparrow + H_2O$$

$$K_2Cr_2O_7 + H_2SO_4 + 3SO_2 \longrightarrow K_2SO_4 + \underset{\substack{\text{Chromium sulphate}\\ \text{(Green)}}}{Cr_2(SO_4)_3} + H_2O$$

4. ***Nitrites.*** On treating the solid nitrite with dil. H_2SO_4, nitric oxide (NO) gas is evolved which readily gives dense brown fumes of NO_2 with oxygen of the air.

$$KNO_2 + H_2SO_4 \longrightarrow KHSO_4 + HNO_2\,] \times 3$$

$$3HNO_2 \longrightarrow HNO_3 + H_2O + 2NO$$

$$\overline{3KNO_2 + 3H_2SO_4 \longrightarrow 3KHSO_4 + HNO_3 + H_2O + 2NO}$$

$$\underset{\text{Colourless}}{2NO} + O_2 \longrightarrow \underset{\substack{\text{Brown}\\ \text{fumes}}}{2NO_2}$$

12.08. POTASSIUM PERMANGANATE TEST

To a pinch of salt in test tube add about 2 ml of dilute sulphuric acid. Boil off any gas evolved, add little more of dilute acid and then potassium permanganate solution dropwise. Note the changes as given in Table 12.8. This test helps in detection of Cl^-, Br^-, I^-, $C_2O_4{}^{2-}$ and Fe^{2+} radicals.

Table 12.8. Potassium Permanganate Test

Observations	***Inference***
1. Potassium permanganate decolourised without the evolution of any gas.	Presence of Fe^{2+} salts.
2. Potassium permanganate decolourised:	
(a) In cold	
(*i*) With the evolution of chlorine.	Cl^-
(*ii*) With the evolution of bromine.	Br^-
(*iii*) With the evolution of iodine.	I^-
(b) On warming	
With evolution of CO_2	$C_2O_4{}^{2-}$
3. $KMnO_4$ not decolourised.	Absence of Cl^-, Br^-, I^-, $C_2O_4{}^{2-}$ and Fe^{2+}.

Notes:

1. *As sulphides are oxidised by $KMnO_4$ so they have to be completely decomposed by heating with dilute sulphuric acid before this test is performed.*
2. *Potassium permanganate oxidises Fe^{2+} salts in cold. Dil. H_2SO_4 acid is added to the salt and heated till sulphides, sulphites and nitrites are completely decomposed. Then $KMnO_4$ is added dropwise to cold solution.*

Chemical Reactions Involved

$$2KMnO_4 + 3H_2SO_4 \longrightarrow K_2SO_4 + 2MnSO_4 + 3H_2O + 5[O]$$

1. *Ferrous salts:*

$$2FeSO_4 + H_2SO_4 + [O] \longrightarrow Fe_2(SO_4)_3 + H_2O$$

2. *Chlorides:*

$$NaCl + H_2SO_4 \longrightarrow NaHSO_4 + HCl$$
$$2HCl + [O] \longrightarrow H_2O + Cl_2\uparrow$$

3. *Bromides:*

$$NaBr + H_2SO_4 \longrightarrow NaHSO_4 + HBr$$
$$2HBr + [O] \longrightarrow H_2O + Br_2\uparrow$$

4. *Iodides:*

$$NaI + H_2SO_4 \longrightarrow NaHSO_4 + HI$$
$$2HI + [O] \longrightarrow H_2O + I_2\uparrow$$

5. *Oxalates:*

$$\begin{matrix}COONa\\ |\\ COONa\end{matrix} + H_2SO_4 \longrightarrow \begin{matrix}COOH\\ |\\ COOH\end{matrix} + Na_2SO_4$$

$$\begin{matrix}COOH\\ |\\ COOH\end{matrix} + [O] \longrightarrow 2CO_2\uparrow + H_2O$$

12.09. CONCENTRATED SULPHURIC ACID TEST

This test is performed by treating small quantity of salt with conc. sulphuric acid (2–3 ml) in a test tube. Identify the gas evolved in cold and then on heating. Observe the changes and draw inferences as given in Table 12.9.

Table 12.9. Conc. Sulphuric Acid Test

Observations	***Inference***	
	Gas evolved	***Radical***
1. Colourless gas with pungent smell, white fumes with aqueous ammonia (NH_4OH), white ppt. with $AgNO_3$ solution.	HCl	Cl^-
2. Reddish brown vapours with pungent smell, turns starch paper yellow. It does not turn $FeSO_4$ solution black.	Br_2	Br^-
3. Deep violet vapours with pungent smell, turns starch paper blue. A sublimate is formed on the sides of the tube.	I_2 vapours	I^-
4. Reddish brown gas with pungent smell, turns $FeSO_4$ solution black.	NO_2	NO_3^-
5. Colourless vapours with smell of vinegar, turns blue litmus red.	CH_3COOH vapours,	CH_3COO^-
6. A colourless gas which turns lime water milky and also a gas which burns with pale-bluish flame.	$CO_2 + CO$	$C_2O_4^{2-}$
7. No gas/vapours evolve.	—	Cl^-, Br^-, I^-, $C_2O_4^{2-}$, CH_3COO^- absent

Notes:

1. *If some gas evolves with dilute sulphuric acid, then there is no need for performing conc. sulphuric acid test.*
2. *Some acetates may react with dilute sulphuric acid and give out vapours of acetic acid in dilute sulphuric acid test.*
3. *Do not boil the salt with conc. sulphuric acid. On boiling, the acid may decompose to give SO_2 gas.*
4. *Nitrates give vapours of nitric acid (colourless) when heated with conc. sulphuric acid. When a paper pellet or copper chips is added, dense brown fumes evolve. Paper pellet acts as a reducing agent and reduces nitric acid to NO_2 (Reddish brown gas).*

$$NaNO_3 + H_2SO_4 \longrightarrow NaHSO_4 + HNO_3$$

$$4HNO_3 + \underset{\text{(From paper pellet)}}{C} \longrightarrow 2H_2O + 4NO_2\uparrow + CO_2\uparrow$$

Chemical Reactions Involved in conc. H_2SO_4 Test

1. ***Chlorides***

$$NaCl + H_2SO_4 \longrightarrow \underset{\text{Sod. bisulphate}}{NaHSO_4} + HCl$$

2. ***Bromides***

$$NaBr + H_2SO_4 \longrightarrow NaHSO_4 + HBr$$

$$H_2SO_4 + 2HBr \longrightarrow SO_2 + Br_2 + 2H_2O$$

3. ***Iodides***

$$KI + H_2SO_4 \longrightarrow KHSO_4 + HI$$

$$H_2SO_4 + 2HI \longrightarrow SO_2 + I_2 + 2H_2O$$

4. ***Nitrates***

$$KNO_3 + H_2SO_4 \longrightarrow KHSO_4 + HNO_3$$

$$4HNO_3 + \underset{\text{(Paper pellet)}}{C} \longrightarrow 4NO_2\uparrow + CO_2 + 2H_2O$$

5. ***Acetates***

$$CH_3COONa + H_2SO_4 \longrightarrow NaHSO_4 + \underset{\text{Acetic acid (Vinegar smell)}}{CH_3COOH}$$

6. ***Oxalates***

$$\begin{array}{l} COONa \\ | \\ COONa \end{array} + H_2SO_4 \longrightarrow Na_2SO_4 + CO_2\uparrow + CO\uparrow + H_2O$$

12.10. TESTS FOR INDEPENDENT RADICALS (SO_4^{2-} AND PO_4^{3-})

As already discussed these radicals are not detected by dilute or concentrated H_2SO_4. They are tested individually.

1. Sulphate (SO_4^{2-})

Boil a small amount of salt with dilute HCl in a test tube. Filter the contents, and to the filtrate add few drops of $BaCl_2$ solution. A **white ppt.** insoluble in conc. HCl indicates presence of sulphate.

2. Phosphate (PO_4^{3-})

Add conc. HNO_3 to the salt in a test tube. Boil the contents and add excess of ammonium molybdate solution. **A yellow precipitate** indicates presence of phosphate.

12.11. CONFIRMATION OF ACID RADICALS BY WET TESTS

The acid radical indicated by dil. H_2SO_4 or conc. H_2SO_4 tests is further confirmed by wet tests.

Preparation of Solution of the Salt for Wet Tests of Acid Radicals

The confirmatory tests for acid radicals are performed with the solution of the salt. The solution used for the purpose is any one of the following:

1. **Aqueous solution or 'water extract'.** Shake a little of the salt with water. If the salt dissolves, this aqueous solution obtained is used for the wet tests of acid radical and is called 'water extract' or 'W.E.'.
2. **Sodium carbonate extract.** This is prepared only if the salt is insoluble in water.

   Preparation of Sodium Carbonate Extract. Mix about 1 g of the salt with about 2 g of pure sodium carbonate and boil it for 10–15 minutes with 20–25 ml of distilled water in a small conical flask having a funnel in its mouth (Fig. 12.8). The funnel acts as a condenser. This arrangement prevents the loss of water due to evaporation. Filter the solution, cool it and label it as Sodium Carbonate Extract Or S.E.

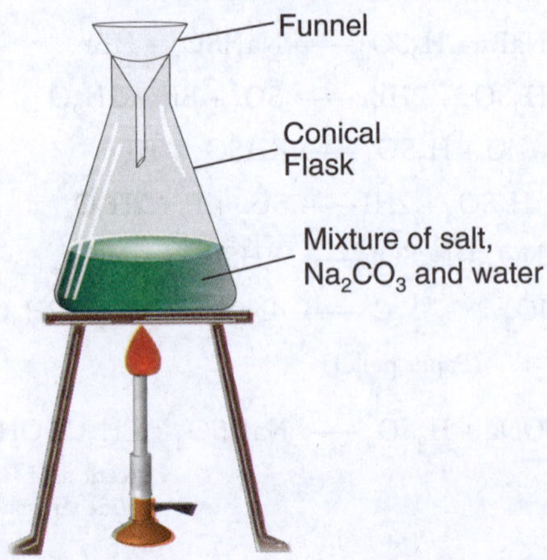

Fig. 12.8. Preparation of sodium carbonate extract.

Alternatively, sodium carbonate extract can be prepared in a test tube. A pinch of salt is mixed with double the amount of sodium carbonate and is boiled with distilled water for sometime. The suspension obtained is filtered. The filtrate is sodium carbonate extract.

Theory of Preparation of Sodium Carbonate Extract.

When the salts are boiled with strong solution of sodium carbonate, double decomposition reaction takes place resulting in the formation of the carbonates of heavy metallic radicals and sodium salts of the acid radicals. The sodium salts of corresponding acid radicals being soluble in water pass into the solution and carbonates of heavy metals are precipitated.

$$ZnS(s) + Na_2CO_3(aq) \longrightarrow ZnCO_3(s) \downarrow + Na_2S(aq)$$

How to Use Sodium Carbonate Extract?

Sodium carbonate extract always contains unreacted sodium carbonate in solution which has to be destroyed before using the extract for various tests. To do this, the extract is acidified with some suitable acid and is boiled to expel carbon dioxide. The selection of acid used for destroying excess Na_2CO_3 depends upon the radical to be identified.

Now we describe in detail the confirmatory tests for various acid radicals discussed so far.

Confirmation of Carbonate, CO_3^{2-}

(Indicated in dilute acid test by occurrence of brisk effervescence and evolution of carbon dioxide).

Confirmation of soluble carbonate	*Confirmation of insoluble carbonate*
If the salt dissolves, soluble carbonate is indicated.	If the salt remains insoluble, the presence of insoluble carbonate is indicated.
1. Dil HCl test To one portion of the solution, add dil. HCl.	To the salt add dil. HCl.
Brisk effervescence and evolution of carbon dioxide which turns lime water milky confirms the presence of soluble carbonate.	**Brisk effervescence** and **evolution of carbon diaxide** which turns lime water milky confirms the presence of **insoluble carbonate.**
2. Magnesium sulphate test To another portion of the solution, add magnesium sulphate solution. Formation of **white precipitate** in the cold confirms the presence of **soluble carbonate.**	

Notes:

1. *Do not use sodium carbonate extract for performing the tests of carbonates because it contains sodium carbonate.*
2. *Perform magnesium sulphate test only in case of soluble carbonates.*

Confirmation of Sulphite, SO_3^{2-}

(Indicated in dilute acid test by the evolution of sulphur dioxide).

Experiment	*Observations*
1. Barium chloride test Take a portion of aqueous solution (or sodium carbonate extract and dil. acetic acid and boil off CO_2). Add barium chloride solution to it. Filter.	A white ppt. is formed.
To a portion of the above ppt. add dil. HCl.	The ppt. dissolves with the evolution of sulphur dioxide.
2. $KMnO_4$ test To a second part of the ppt. from (1) add a few drops of acidified potassium permanganate solution.	The pink colour is discharged.
3. $K_2Cr_2O_7$ test To a portion of aqueous solution or sodium carbonate extract add potassium dichromate solution acidified with dil. H_2SO_4.	A green colour is obtained.

Confirmation of Sulphide, S^{2-}

(Indicated in dilute acid test by the evolution of hydrogen sulphide).

Experiment	*Observations*
1. Sodium nitroprusside test Take a portion of aqueous solution (or sodium carbonate extract) in a test tube and add a few drops of *sodium nitroprusside solution.*	**Purple or violet colouration** is obtained.
2. Lead acetate test To a portion of aqueous solution (or sodium carbonate extract acidified with dil. acetic acid) add lead acetate solution.	A **black ppt.** is obtained.
3. Cadmium carbonate test To a portion of aqueous solution (or *sodium carbonate extract*) add a suspension of *cadmium carbonate in water.*	A **yellow ppt.** is formed.

Confirmation of Nitrite, NO_2^-

(Indicated in dilute acid test by the evolution of brown vapours of nitrogen dioxide)

Experiment	*Observations*
1. Ferrous sulphate test To a portion of aqueous solution, add some dil. acetic acid and ferrous sulphate solution.	A **dark brown or black colouration** is obtained.
2. Starch-iodide test To a portion of aqueous solution add a few drops of dil. H_2SO_4 and a few drops of potassium iodide solution followed by freshly prepared starch solution.	A **blue solution** is obtained.
3. Diphenylamine test To a portion of aqueous solution, add a few drops of diphenylamine.	A **deep blue colouration** is obtained.

Chemical Reactions Involved in the Confirmation of Carbonate, Sulphite, Sulphide and Nitrite

Carbonate (CO_3^{2-})

1. *Reaction with dil. HCl*

Carbonates on reaction with dil. HCl give CO_2 gas which turns lime water milky. In case of soluble carbonates this test is performed with water extract and in case of insoluble carbonates this test is performed with the solid salt.

$$CaCO_3 + 2HCl \longrightarrow CaCl_2 + CO_2 + H_2O$$

$$\underset{\text{Lime water}}{Ca(OH)_2} + CO_2 \longrightarrow \underset{\text{Milkiness}}{CaCO_3} + H_2O$$

2. *Magnesium sulphate test*

This test is performed in case of soluble carbonates only

$$Na_2CO_3 + MgSO_4 \longrightarrow \underset{\text{(White ppt.)}}{MgCO_3 \downarrow} + Na_2SO_4$$

Sulphite (SO_3^{2-})

1. *Barium chloride test*

$$Na_2SO_3 + BaCl_2 \longrightarrow 2NaCl + \underset{\text{(White ppt.)}}{BaSO_3 \downarrow}$$

$$BaSO_3 + 2HCl \longrightarrow BaCl_2 + SO_2 \uparrow + H_2O$$

2. *Potassium permanganate test*

$$2KMnO_4 + 3H_2SO_4 \longrightarrow K_2SO_4 + 2MnSO_4 + 3H_2O + 5[O]$$

$$Na_2SO_3 + H_2SO_4 \longrightarrow Na_2SO_4 + SO_2 + H_2O$$

$$SO_2 + H_2O + [O] \longrightarrow H_2SO_4$$

3. *Potassium dichromate test*

$$\underset{\text{(orange)}}{K_2Cr_2O_7} + 4H_2SO_4 \longrightarrow K_2SO_4 + \underset{\text{(green)}}{Cr_2(SO_4)_3} + 4H_2O + 3[O]$$

$$Na_2SO_3 + [O] \longrightarrow Na_2SO_4$$

Sulphide (S^{2-})

1. *Sod. nitroprusside test*

$$Na_2S + \underset{\text{Sod. nitroprusside}}{Na_2[Fe(CN)_5NO]} \longrightarrow \underset{\text{(Violet or Purple colour)}}{Na_4[Fe(CN)_5NOS]}$$

2. *Lead acetate test*

$$Na_2S + (CH_3COO)_2Pb \longrightarrow \underset{\text{Black ppt.}}{PbS \downarrow} + 2CH_3COONa$$

3. *Cadmium carbonate test*

$$Na_2S + CdCO_3 \longrightarrow \underset{\text{Yellow ppt.}}{CdS \downarrow} + Na_2CO_3$$

Nitrite (NO_2^-)

1. *Ferrous sulphate test*

$$KNO_2 + CH_3COOH \longrightarrow CH_3COOK + HNO_2$$

$$3HNO_2 \longrightarrow HNO_3 + 2NO + H_2O$$

$$FeSO_4 + NO \longrightarrow \underset{\text{(Dark brown)}}{FeSO_4 . NO}$$

2. *Potassium iodide test*

$$KNO_2 + CH_3COOH \longrightarrow HNO_2 + CH_3COOK$$

$$2HNO_2 + 2KI + CH_3COOH \longrightarrow 2CH_3COOK + 2NO + 2H_2O + I_2$$

$$I_2 + \text{Starch} \longrightarrow \text{Blue complex}$$

I_2 turns starch paper blue.

Confirmation of Chloride, Cl^-

(No action with dilute H_2SO_4 but decomposed by conc. H_2SO_4 with the evolution of HCl gas).

Experiment	*Observations*
1. Silver nitrate test Acidify a portion of aqueous solution (or sodium carbonate extract) with dil. HNO_3. Boil for some time, cool and add silver nitrate solution.	A **white ppt.** is formed which is soluble in ammonium hydroxide.
2. Manganese dioxide test Heat a pinch of the salt with a small quantity of manganese dioxide and conc. H_2SO_4.	Evolution of **greenish yellow gas** having a pungent irritating smell. It turns moist starch-iodide paper blue.
3. Chromyl chloride test Mix a small quantity of the salt with a small amount of powdered potassium dichromate. Take the mixture in a test tube and add conc. H_2SO_4. Heat the tube and pass the red vapours evolved into the gas detector containing NaOH solution. To the yellow solution thus obtained, add dil. acetic acid and lead acetate solution.	A **yellow ppt.** is formed.

Confirmation of Bromide, Br^-

(No action with dilute H_2SO_4 but decomposed by conc. H_2SO_4 with the evolution of bromine vapours).

Experiment	*Observations*
1. Silver nitrate test Acidify a portion of aqueous solution (or sodium carbonate extract) with dil. HNO_3. Boil, cool and add silver nitrate solution.	A **light yellow ppt.** is obtained which is partially soluble in NH_4OH.
2. Manganese dioxide test Heat a small quantity of the salt with solid manganese dioxide and conc. H_2SO_4.	Evolution of **yellowish brown vapours** of bromine which turn starch paper yellow.
3. Chlorine water test Acidify a portion of aqueous solution (or sodium carbonate extract) with dil. HCl and add 1–2 ml of carbon disulphide and then chlorine water. Shake vigorously and allow to stand.	Carbon disulphide layer acquires orange colouration.

Note: *Chlorine water is prepared by adding dropwise conc. HCl to a small volume of $KMnO_4$ solution till the pink colour is just discharged, the resulting solution is chlorine water.*

Confirmation of Iodide, I^-

(No action with dilute H_2SO_4 but decomposed by conc. H_2SO_4 with the evolution of vapours of iodine).

Experiment	Observations
1. Silver nitrate test Acidify a portion of aqueous solution (or sodium carbonate extract) with dil. HNO_3. Boil, cool and add silver nitrate solution.	A **yellow ppt.** is formed which is insoluble in NH_4OH.
2. Manganese dioxide test Heat a small quantity of the salt with a little manganese dioxide and conc. H_2SO_4.	Evolution of **violet vapours of iodine** which turn starch paper blue.
3. Chlorine water test Acidify a part of the aqueous solution (or sodium carbonate extract) with dil. HCl, add 1-2 ml of carbon disulphide and then chlorine water. Shake vigorously and allow to stand.	Carbon disulphide layer acquires a violet colouration.

Confirmation of Nitrate, NO_3^-

(No action with dilute acids but decomposed by conc. H_2SO_4 with the evolution of brown vapours of nitrogen peroxide).

Experiment	Observations
1. Copper chips test Heat a small quantity of the original salt with concentrated sulphuric acid and a few copper chips.	**Dark brown fumes** of nitrogen dioxide are evolved.
2. Ring Test Add a small quantity of freshly prepared solution of ferrous sulphate to a part of the aqueous solution and then pour concentrated sulphuric acid slowly along the sides of the test tube as shown in Fig. 12.9. 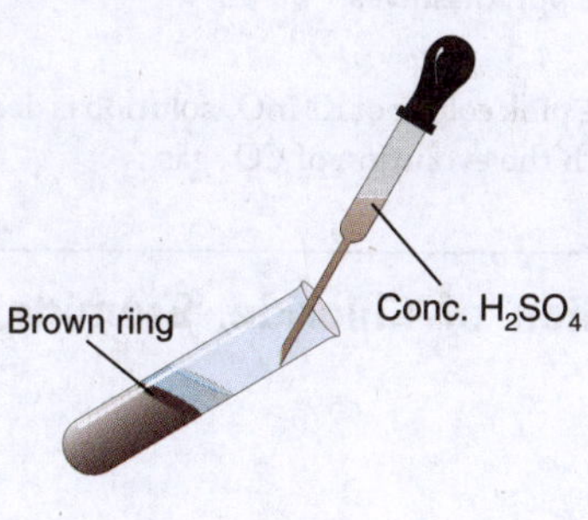 Fig. 12.9	A **dark brown ring** is formed at the junction of the layers of the acid and the solution.

Confirmation of Acetate, CH_3COO^-

(No action with dilute acids but decomposed by conc. H_2SO_4 with the evolution of CH_3COOH vapours)

Experiment	Observations
1. Oxalic acid test	
Take a small quantity of the salt on a watch glass. Mix it with solid oxalic acid. Prepare paste of it with a few drops of water. Rub the paste and smell.	Smell like that of vinegar.
2. Ester test	
Take a small quantity of the salt in a test-tube. Add conc. H_2SO_4 (2 ml) and heat. Now add ethyl alcohol (1 ml). Shake and pour the contents of the tube in a beaker full of water. Stir.	Pleasant fruity smell of ester.
3. Ferric chloride test	
Take water extract of the salt. Add *neutral* ferric chloride solution. Filter and divide the filtrate into two portions.	Reddish coloured filtrate.
(*i*) To one part, add dil. HCl.	Reddish colour disappears.
(*ii*) To second part, add water and boil.	Reddish brown ppt.

Note: *In order to prepare* ***neutral ferric chloride solution,*** *take about 5 ml of ferric chloride solution and add ammonium hydroxide solution dropwise until a small but permanent precipitate of ferric hydroxide is obtained. Boil the solution and remove the precipitate (if any) by centrifugation or filtration. The clear solution thus obtained is neutral ferric chloride solution.*

Confirmation of Oxalate, $C_2O_4^{2-}$

(No action with dilute acids but decomposed by conc. H_2SO_4 with the evolution of CO_2 and CO gas)

Experiment	Observations
1. Calcium chloride test	
Take water extract (or soda extract if salt is insoluble in water). Acidify with dil acetic acid and boil off CO_2. Add calcium chloride solution.	A **white ppt.** is formed.
Add dil. HNO_3 to the white ppt and warm.	The ppt. dissolves.
2. Potassium permanganate test	
Take a pinch of the salt in test tube and add dil sulphuric acid. Warm to 60–70°C and add 2–3 drops of $KMnO_4$ solution.	The pink colour of $KMnO_4$ solution is decolourized with the evolution of CO_2 gas.

Chemical Reactions Involved in the Confirmation of Chloride, Bromide, Iodide, Acetate and Oxalate

Chloride (Cl^-)

1. *Silver nitrate test*

$$NaCl + AgNO_3 \longrightarrow \underset{\text{White ppt.}}{AgCl\downarrow} + NaNO_3$$

$$AgCl + 2NH_4OH \longrightarrow \underset{\text{Soluble complex}}{[Ag(NH_3)_2]\,Cl} + 2H_2O$$

2. *Manganese dioxide test*

$$2NaCl + MnO_2 + 3H_2SO_4 \longrightarrow 2NaHSO_4 + MnSO_4 + 2H_2O + Cl_2\uparrow$$

3. *Chromyl chloride test*

$$4NaCl + K_2Cr_2O_7 + 6H_2SO_4 \longrightarrow 4NaHSO_4 + 2KHSO_4 + \underset{\text{Chromyl chloride}}{2CrO_2Cl_2} + 3H_2O$$

$$CrO_2Cl_2 + 4NaOH \longrightarrow \underset{\text{Sod. chromate}}{Na_2CrO_4} + 2H_2O + 2NaCl$$

$$Na_2CrO_4 + (CH_3COO)_2Pb \longrightarrow \underset{\substack{\text{Lead chromate}\\ \text{(Yellow ppt.)}}}{PbCrO_4\downarrow} + 2CH_3COONa$$

Bromide (Br^-)

1. *Silver nitrate test*

$$KBr + AgNO_3 \longrightarrow KNO_3 + \underset{\text{(Pale yellow ppt.)}}{AgBr\downarrow}$$

Pale yellow ppt. of silver bromide are sparingly soluble in ammonium hydroxide.

2. *Manganese dioxide test*

$$2KBr + MnO_2 + 3H_2SO_4 \longrightarrow 2KHSO_4 + MnSO_4 + 2H_2O + Br_2$$

3. *Chlorine water test*

$$2KBr + Cl_2 \longrightarrow 2KCl + Br_2$$

Bromine being soluble in CCl_4 imparts an orange colour to the CCl_4 layer.

Iodide (I^-)

1. *Silver nitrate test*

$$KI + AgNO_3 \longrightarrow KNO_3 + \underset{\text{(yellow ppt.)}}{AgI}$$

2. *Manganese dioxide test*

$$2KI + MnO_2 + 3H_2SO_4 \longrightarrow 2KHSO_4 + MnSO_4 + 2H_2O + I_2$$

3. *Chlorine water test*

$$2KI + Cl_2 \longrightarrow 2KCl + I_2$$

Iodine being soluble in CCl_4 imparts a violet colour to the CCl_4 layer.

Nitrate (NO_3^-)

1. *Copper test*

$$2KNO_3 + H_2SO_4 \longrightarrow K_2SO_4 + 2HNO_3$$

$$4HNO_3 + Cu \longrightarrow Cu(NO_3)_2 + \underset{\text{(Reddish brown)}}{2NO_2} + 2H_2O$$

2. *Ring test*

$$KNO_3 + H_2SO_4 \longrightarrow KHSO_4 + HNO_3$$

$$6FeSO_4 + 3H_2SO_4 + 2HNO_3 \longrightarrow 3Fe_2(SO_4)_3 + 4H_2O + 2NO$$

$$FeSO_4 + NO + 5H_2O \longrightarrow \underset{\text{(Brown ring)}}{[Fe(NO)(H_2O)_5]SO_4}$$

Acetate (CH_3COO^-)

1. *Oxalic acid test*

$$\begin{matrix} COOH \\ | \\ COOH \end{matrix} + 2CH_3COONa \longrightarrow \begin{matrix} COONa \\ | \\ COONa \end{matrix} + \underset{\text{Acetic acid (Vinegar smell)}}{2CH_3COOH}$$

2. *Ester test*

$$2CH_3COONa + H_2SO_4 \longrightarrow Na_2SO_4 + 2CH_3COOH$$

$$CH_3COOH + C_2H_5OH \longrightarrow \underset{\substack{\text{Ethyl acetate} \\ \text{(Fruity smell)}}}{CH_3COOC_2H_5} + H_2O$$

3. *Ferric chloride test*

$$3CH_3COONa + FeCl_3 \longrightarrow (CH_3COO)_3Fe + 3NaCl$$

$$(CH_3COO)_3Fe + 2H_2O \longrightarrow \underset{\text{Reddish brown ppt.}}{(CH_3COO)(OH)_2\,Fe\downarrow} + 2CH_3COOH$$

Oxalate ($C_2O_4^{2-}$)

1. *Calcium chloride test*

$$(NH_4)_2C_2O_4 + CaCl_2 \longrightarrow \underset{\text{White ppt.}}{CaC_2O_4\downarrow} + 2NH_4Cl$$

2. *Potassium permanganate test*

$$2KMnO_4 + 3H_2SO_4 + 5\begin{matrix} COOH \\ | \\ COOH \end{matrix} \longrightarrow K_2SO_4 + 2MnSO_4 + 10CO_2 + 8H_2O$$

Confirmation of Sulphate, SO_4^{2-}

(Not indicated in dilute and concentrated H_2SO_4 acid tests).

Experiment	*Observations*
1. Barium chloride test To a part of the aqueous solution of the salt add barium chloride solution.	A **white ppt.** is formed which is insoluble in dil HCl.
2. Match stick test Mix a small amount of the salt with sodium carbonate and a little powdered charcoal so as to get a paste. Take some of this paste on one end of a wooden splinter and heat in the reducing flame till the mass fuses. Dip the fused mass into sodium nitroprusside solution taken in a china dish.	**Violet streaks** are produced.
3. Lead acetate test To a part of aqueous solution of the salt add lead acetate solution.	A **white ppt.** is formed which is soluble in excess of hot ammonium acetate solution.

Confirmation of Phosphate, PO_4^{3-}

(Not indicated in dilute and concentrated H_2SO_4 acid test).

Experiment	*Observations*
1. Ammonium molybdate test To the aqueous solution or sodium carbonate extract (or the original salt) add concentrated nitric acid and boil. Add ammonium molybdate solution in excess and again boil.	A **deep yellow ppt.** or colouration is obtained.
2. Magnesia mixture test Take a portion of aqueous solution (or a part of sodium carbonate extract, add hydrochloric acid to acidify it and boil off CO_2). Add magnesia mixture (to prepare it, add solid NH_4Cl to magnesium chloride solution. Boil, cool and add NH_4OH till a strong smell of ammonia is obtained) and allow to stand.	A **white ppt.** is obtained.

Chemical Reactions Involved in the Confirmation of SO_4^{2-} and PO_4^{3-}

Sulphate (SO_4^{2-})

1. *Barium chloride test*

$$Na_2SO_4 + BaCl_2 \longrightarrow \underset{\text{white}}{BaSO_4} \downarrow + 2NaCl$$

2. *Match-stick test*

$$Na_2SO_4 + BaCl_2 \longrightarrow BaSO_4 \downarrow + 2NaCl$$

$$BaSO_4 + Na_2CO_3 \longrightarrow Na_2SO_4 + BaCO_3$$

$$Na_2SO_4 + 4C \longrightarrow Na_2S + 4CO$$

$$Na_2S + Na_2[Fe(CN)_5NO] \longrightarrow \underset{\text{Purple}}{Na_4[Fe(CN)_5NOS]}$$

3. *Lead acetate test*

$$Na_2SO_4 + Pb(CH_3COO)_2 \longrightarrow PbSO_4 \downarrow + 2CH_3COONa$$

Phosphate (PO_4^{3-})

1. *Ammonium molybdate test*

$$K_3PO_4 + \underset{\text{Amm. molybdate}}{3(NH_4)_2 MoO_4} \longrightarrow 2(NH_4)_3PO_4 + \underset{\text{Pot. molybdate}}{3K_2MoO_4}$$

$$K_2MoO_4 + 2HNO_3 \longrightarrow \underset{\text{Molybdic acid}}{H_2MoO_4} + 2KNO_3$$

$$H_2MoO_4 \longrightarrow MoO_3 + H_2O$$

$$(NH_4)_3PO_4 + 12MoO_3 + 6H_2O \longrightarrow \underset{\substack{\text{Amm. phosphate molybdate} \\ \text{(yellow ppt).}}}{(NH_4)_3PO_4 . 12MoO_3 . 6H_2O} \downarrow$$

2. *Magnesia mixture test*

$$\underset{\text{Disodium hydrogen phosphate}}{Na_2HPO_4} + MgCl_2 + NH_4OH \longrightarrow \underset{\text{Mag. amm. phosphate (white ppt.)}}{Mg(NH_4)PO_4 \downarrow} + 2NaCl + H_2O.$$

WET TESTS FOR BASIC RADICALS (CATIONS)

Preliminary tests such as dry heating test, charcoal cavity test, flame test and borax bead test may give us some indication about the cation present in the salt. However, the cation is finally detected and confirmed through a systematic analysis involving wet tests. For the sake of qualitative analysis the cations are classified into the following groups (Table 12.10).

Table 12.10. Classification of Cations

Group	***Cations***
Group zero	NH_4^+
Group I	Pb^{2+}
Group II	Pb^{2+}, Cu^{2+}, As^{3+}
Group III	Fe^{3+}
Group IV	Co^{2+}, Ni^{2+}, Mn^{2+}, Zn^{2+}
Group V	Ba^{2+}, Sr^{2+}, Ca^{2+}
Group VI	Mg^{2+}

Before carrying out the wet tests for the analysis of cation, the salt has to be dissolved in some suitable solvent to prepare its solution.

12.12. PREPARATION OF SOLUTION FOR WET TESTS OF BASIC RADICALS

The very first essential step is to prepare a clear and transparent solution of the salt under investigation. For this purpose, the under noted solvents are tried one after another in a systematic order. In case the salt does not dissolve in a particular solvent even on heating, try the next solvent. The following solvents are tried:

(*i*) *Distilled water* (cold or hot).

(*ii*) *Dilute HCl* (cold or hot).

(*iii*) *Conc. HCl* (cold or hot).

PROCEDURE FOR THE PREPARATION OF SOLUTION

1. Take a small quantity of the given salt and add some distilled water. Shake the contents. If the salt does not dissolve, heat the contents and observe whether the salt completely dissolves or not.
2. If the salt does not dissolve in distilled water (cold as well as hot), take the first quantity of the salt in a clean test tube and try to dissolve first in cold dil. HCl and then in hot dil. HCl.
3. If the salt does not dissolve in distilled water as well as dil. HCl, then try to dissolve it in conc. HCl, first in cold and then in hot.

   The clear solution thus obtained is labelled as **Original Solution (O.S.).**

Important Notes:

1. In case some gas is evolved during the preparation of solution, let the reaction cease. Gas must be completely expelled by heating.
2. In case solution is prepared in dilute HCl, group I is absent. Proceed with group II.
3. If the salt is soluble in hot water, and on cooling white precipitates appear, lead chloride is indicated.
4. It is necessary to dilute the solution if it is made in concentrated acid before proceeding with the analysis.

The following table will help the students in the choice of a suitable solvent:

Solvent	***Salts which dissolve***
1. Cold water	(*a*) All NH_4^+, Na^+ and K^+ salts. (*b*) All nitrites, nitrates and acetates. (*c*) Most of the sulphates except those of Pb, Ba, Ca, Sr. (*d*) All chlorides except that of lead.
2. Hot water	Lead chloride, lead nitrate.
3. Dil. HCl	All carbonates which do not dissolve in water *i.e.*, Carbonates of Ca, Ba, Sr, Mg, Zn, Al, Cu, Ni, Mn, Fe etc., but not of Pb.

The separation of cations into various groups by making use of suitable reagents (known as group reagents) is based on the differences in chemical properties of cations. For example, if hydrochloric acid is added to a solution containing all cations, only the chlorides of lead, silver and mercury (ous) will precipitate, since all other chlorides are soluble. Thus, these cations form a group of ions which may be precipitated from solution by addition of group reagent HCl. Similarly, H_2S is a group reagent for group II. The following Table 12.11 clearly shows the group reagents for different groups and the form in which cations of the particular group are precipitated out.

Table 12.11. Group Reagents

Group	***Group reagent***	***Cations***	***Form in which cations are precipitated***
Group zero	No	—	—
Group I	Dilute HCl	Pb^{2+}	Chlorides
Group II	H_2S in the presence of dilute HCl	Pb^{2+}, Cu^{2+}, As^{3+}	Sulphides
Group III	NH_4OH in the presence of NH_4Cl	Fe^{3+}, Al^{3+}	Hydroxides
Group IV	H_2S in the presence of NH_4OH	Co^{2+}, Ni^{2+}, Mn^{2+}, Zn^{2+}	Sulphides
Group V	$(NH_4)_2$ CO_3 in the presence of NH_4OH	Ca^{2+}, Ba^{2+}, Sr^{2+}	Carbonates
Group VI	No	Mg^{2+}	—

12.13. THEORY OF PRECIPITATION OF DIFFERENT GROUPS

The classification of cations into different groups in the inorganic qualitative analysis is based upon the knowledge of solubility products of salts of these basic radicals. For example, chlorides of Hg_2^{2+}, Pb^{2+} and Ag^+ have very low solubility products. On the basis of this knowledge these radicals are grouped together in group-I and are precipitated as their chlorides by adding dilute HCl to their solutions. For adjusting the conditions for precipitation, another concept called *common ion effect* plays very important role. Before we consider the precipitation of radicals of other groups, let us discuss in brief the concept of common ion effect.

COMMON ION EFFECT

Weak acids and weak bases are ionised only to small extent in their aqueous solutions. In their solutions, unionised molecules are in dynamic equilibrium with ions. The degree of ionisation of a weak electrolyte (weak acid or weak base) is further suppressed if some strong electrolyte which can furnish some ion common with the ions furnished by weak electrolyte, is added to its solution. This effect is called **common ion effect.** For example, degree of ionisation of NH_4OH (a weak base) is suppressed by the addition of NH_4Cl (a strong electrolyte). The ionisation of NH_4OH and NH_4Cl in solution is represented as follows:

$$NH_4OH\,(aq) \rightleftharpoons \boxed{NH_4^+\,(aq)} + OH^-\,(aq) \;\ldots\; \textit{weakly ionised} \qquad \ldots(12.1)$$

$$NH_4Cl \longrightarrow \boxed{NH_4^+\,(aq)} + Cl^-\,(aq) \;\ldots\; \textit{strongly ionised} \qquad \ldots(12.2)$$

Common ion

Due to the addition of NH_4Cl, which is strongly ionised in the solution, concentration of NH_4^+ ions increases in the solution. Therefore, according to Le-Chatelier's principle equilibrium in equation (12.1) shifts in the backward direction in favour of unionised NH_4OH. In this way, addition of NH_4Cl suppresses the degree of ionisation of NH_4OH. Thus, the concentration of OH^- ions in the solution is considerably reduced and the weak base NH_4OH becomes a still weaker base.

The suppression of the degree of ionisation of a weak electrolyte (weak acid or weak base) by the addition of some strong electrolyte having a common ion, is called the ***common ion effect.***

Application of concept of common ion effect in the qualitative analysis is illustrated as follows:

The cations of **group II (Pb^{2+}, Cu^{2+}, As^{3+})** are precipitated as their sulphides. Solubility products of sulphides of group II radicals are very low. Therefore, even with low concentration of S^{2-} ions, the ionic products (Q_{sp}) exceed the value of their solubility products (K_{sp}) and the radicals of group II get precipitated. The low concentration of S^{2-} ions is obtained by passing H_2S gas through the solution of the salts in the presence of dil. HCl which suppresses degree of ionisation of H_2S by common ion effect.

$$H_2S \rightleftharpoons \boxed{2H^+} + S^{2-} \qquad \ldots(12.3)$$

$$HCl \longrightarrow \boxed{H^+} + Cl \qquad \ldots(12.4)$$

Common ion

It is necessary to suppress the concentration of S^{2-} ions, otherwise radicals of group IV will also get precipitated along with group II radicals.

Radicals of **group IV (Ni^{2+}, Co^{2+}, Mn^{2+}, Zn^{2+})** are also precipitated as their sulphides. But solubility products of their sulphides are quite high. In order that ionic products exceed solubility products, concentration of S^{2-} ions should be high in this case. High concentration of sulphide ions is achieved by passing H_2S gas through the solutions of the salts in the presence of NH_4OH. Hydroxyl ions from NH_4OH combine with H^+ ions from H_2S. Due to the removal of H^+ ions the equilibrium of H_2S shifts in favour of ionised form.

$$H_2S \rightleftharpoons 2H^+ + S^{2-}$$

$$NH_4OH \rightleftharpoons NH_4^+ + OH^-$$

$$H^+ + OH^- \rightleftharpoons H_2O$$

Hence, concentration of S^{2-} ions increases. With this increased concentration of S^{2-} ions ionic products exceed solubility products and radicals of group IV get precipitated.

Radicals of **group III (Fe^{3+}, Al^{3+})** are precipitated as their hydroxides by NH_4OH in the presence of NH_4Cl. The purpose of NH_4Cl is to suppress the degree of ionisation of NH_4OH by common ion effect in order to decrease the concentration of OH^- ions.

$$NH_4OH \rightleftharpoons \boxed{NH_4^+} + OH^-$$
$$NH_4Cl \longrightarrow \boxed{NH_4^+} + Cl^-$$

Common ion

The solubility products of hydroxides of group III radicals are quite low. Therefore, even with this suppressed concentration of OH^- ions their ionic products exceed solubility products and hence they get precipitated. If the concentration of OH^- ions is not suppressed, the radicals of groups IV, V and Mg^{2+} will also be precipitated along with radicals of group III.

Radicals of **group V (Ba^{2+}, Sr^{2+}, Ca^{2+})** are precipitated as their carbonates by the addition of $(NH_4)_2CO_3$ in the presence of NH_4Cl and NH_4OH. NH_4Cl suppresses the degree of ionisation of $(NH_4)_2CO_3$ by common ion effect and hence decreases the concentration of CO_3^{2-} ions.

$$(NH_4)_2CO_3 \rightleftharpoons \boxed{2NH_4^+} + CO_3^{2-}$$
$$NH_4Cl \longrightarrow \boxed{NH_4^+} + Cl^-$$

Common ion

But solubility products of carbonates of group V radicals are quite low and hence even with the suppressed concentration of CO_3^{2-} ions their ionic products exceed solubility products and they get precipitated whereas Mg^{2+} and other radicals of group VI having relatively high solubility products are not precipitated.

12.14. ANALYSIS OF GROUP—ZERO (NH_4^+)

This group includes NH_4^+ cation. During the analysis of cations NH_4Cl and NH_4OH are added in many steps. Therefore, NH_4^+ ion is detected in the beginning using solid salt.

PROCEDURE

The solid salt is heated with concentrated solution of sodium hydroxide. In case, ammonia gas evolves, NH_4^+ is present. Evolution of ammonia gas is **confirmed** by the following tests:

1. Characteristic **ammoniacal smell**.
2. The gas gives **white fumes** when a glass rod dipped in dil. HCl is brought near the mouth of the test tube.
3. When the gas is passed through Nessler's reagent, it would give brown ppt. in case of NH_3.

Chemical Reactions Involved in Group-Zero Analysis

$$NH_4Cl + NaOH \xrightarrow{\Delta} NaCl + H_2O + NH_3\uparrow$$
$$NH_3 + HCl \longrightarrow \underset{\text{White fumes}}{NH_4Cl}$$

Nessler's Reagent Test

$$\underset{\text{Nessler's Reagent}}{2K_2[HgI_4]} + NH_3 + 3KOH \longrightarrow \underset{\text{Brown ppt.}}{H_2N.HgO.HgI\downarrow} + 7KI + 2H_2O$$

12.15. ANALYSIS OF GROUP I (SILVER GROUP)

This group includes Pb^{2+}, Ag^{+} and Hg_2^{2+}. But in the present context, we shall study only Pb^{2+}. **Group reagent for this group is dil. hydrochloric acid.**

PROCEDURE

1. To the original solution add dil. hydrochloric acid. If a **white precipitate** is formed, first group (Pb^{2+}) is present.
2. Filter and wash the ppt. with distilled water and examine as in Table 12.12.

Table 12.12. Analysis of Group I (Pb^{2+})

Boil the white precipitate with 5–10 ml of water. Precipitate dissolves. Divide the solution obtained into three parts. *Confirmation*
1. **Cool one part of the solution under tap.** *White crystalline ppt.* separate out.
2. **Potassium iodide test.** To the second part of the solution, add KI solution—*yellow ppt.*
3. **Potassium chromate test.** To the third part of the solution add K_2CrO_4 solution— *yellow ppt.*

Notes:

1. *If the original solution is prepared in cold dilute hydrochloric acid, first group is absent.*
2. *If the original solution is prepared in conc. hydrochloric acid, simply add water. White ppt. shows the presence of first group.*

Chemical Reactions Involved in Group I Analysis

The addition of HCl to the solution will precipitate Pb^{2+} as chloride

$$Pb(NO_3)_2 + 2HCl \longrightarrow \underset{\text{White ppt.}}{PbCl_2 \downarrow} + 2HNO_3$$

When the white ppt. is boiled with water, the precipitates dissolve because the $PbCl_2$ is soluble in hot water.

Confirmatory tests:

1. On cooling, precipitates settle down as $PbCl_2$ is less soluble in cold water.
2. *Potassium iodide test*

$$\underset{\text{(Hot solution)}}{PbCl_2} + 2KI \longrightarrow \underset{\text{Yellow ppt.}}{PbI_2 \downarrow} + 2KCl$$

3. *Potassium chromate test*

$$\underset{\text{(Hot solution)}}{PbCl_2} + K_2CrO_4 \longrightarrow \underset{\text{Yellow ppt.}}{PbCrO_4 \downarrow} + 2KCl.$$

12.16. ANALYSIS OF GROUP II (COPPER GROUP)

This group includes $\mathbf{Pb^{2+}}$ and $\mathbf{Cu^{2+}}$ in IIA group and $\mathbf{As^{3+}}$ in IIB group. These are precipitated as their sulphides. If group I is absent, the tests for radicals of group II are carried out. **Group reagent for this group is H_2S gas in the presence of dil. HCl.**

PROCEDURE

Take about 2 ml of the original solution in a test tube. Make it acidic with dil. HCl and warm the contents. Through this solution pass H_2S gas from the Kipp's apparatus by turning the stop cock as shown in

Fig. 12.10. Formation of the black or yellow precipitates indicates the presence of group II radical. If this is observed, pass more of H_2S gas to ensure complete precipitation of the radical sulphide. Centrifuge and separate the precipitates.

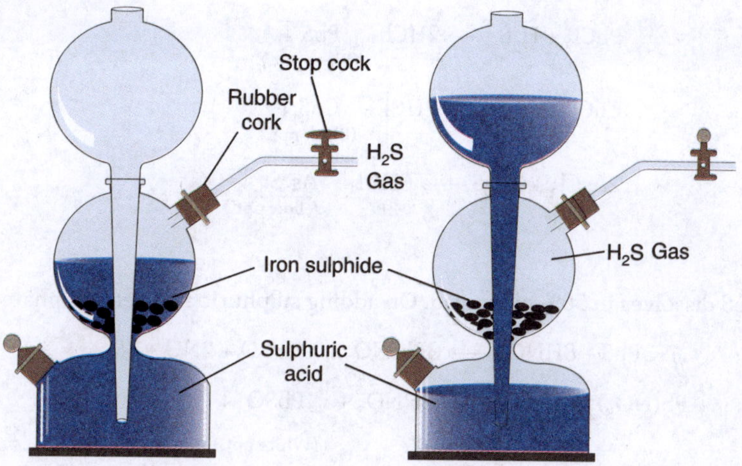

Fig. 12.10. Kipp's apparatus for H_2S gas.

Identification of IIA and IIB Groups. Note the colour of the precipitate. If the precipitate is black in colour, it indicates Pb^{2+} or Cu^{2+}. If the colour of precipitate is yellow this indicates As^{3+}.

Table 12.13. Analysis of Group II

Black ppt. **(Pb^{2+} or Cu^{2+})** *Heat the black ppt. with minimum quantity (1-2 ml) of 50% HNO_3, ppt. dissolves. To one part of the above solution, add dil. H_2SO_4 and alcohol.*		***Yellow ppt.*** **As^{3+}**
White ppt. (Pb^{2+}) **Confirmation** Dissolve the ppt. in hot ammonium acetate solution. Divide the solution into two parts : **1. Potassium iodide test** To one part add pot. iodide solution. **Yellow ppt.** is formed. The ppt. dissolves in boiling water and on cooling recrystallises. **2. Potassium chromate test** To another part add pot. chromate solution. **Yellow ppt.** is formed which dissolves in NaOH solution.	*No white ppt.* *To rest of the solution add NH_4OH in excess* Blue coloured solution (Cu^{2+}) **Confirmation** **1. Potassium ferrocyanide test** To one part of the blue solution add acetic acid and pot. ferrocyanide solution. **A chocolate brown ppt.** is formed. **2. Potassium iodide test** To another part add acetic acid and pot. iodide solution. A **white ppt.** is formed in brown coloured solution.	**Confirmation** Dissolve the yellow ppt. in conc. HNO_3 and divide it into two parts. **1. Ammonium molybdate test.** To a part of the solution, add ammonium molybdate solution and heat—**A yellow ppt.** **2. Magnesia mixture test.** Make the second part of the solution alkaline with NH_4OH solution and add magnesia mixture (contains solutions of $MgSO_4$, NH_4Cl and NH_4OH mixed in equal volumes)—**A white ppt.**

Chemical Reactions Involved in the Analysis of Group II

Passing of H_2S gas through the acidified original solution will precipitate the radicals Pb^{2+}, Cu^{2+} and As^{3+} as their sulphides.

$$PbCl_2 + H_2S \longrightarrow 2HCl + \underset{\text{(Black ppt.)}}{PbS \downarrow}$$

$$CuCl_2 + H_2S \longrightarrow 2HCl + \underset{\text{(Black ppt.)}}{CuS \downarrow}$$

$$2AsCl_3 + 3H_2S \longrightarrow 6HCl + \underset{\text{(Yellow ppt.)}}{As_2S_2 \downarrow}$$

Lead (Pb^{2+})

Black ppt. of PbS dissolves in 50% nitric acid. On adding sulphuric acid, lead sulphate precipitates.

$$3PbS + 8HNO_3 \longrightarrow 3Pb(NO_3)_2 + 4H_2O + 2NO + 3S$$

$$Pb(NO_3)_2 + H_2SO_4 \longrightarrow 2HNO_3 + \underset{\text{(White ppt.)}}{PbSO_4 \downarrow}$$

1. *Potassium iodide test:*

$$Pb(NO_3)_2 + 2KI \longrightarrow 2KNO_3 + \underset{\text{(Yellow ppt.)}}{PbI_2 \downarrow}$$

2. *Potassium chromate test:*

$$Pb(NO_3)_2 + K_2CrO_4 \longrightarrow 2KNO_3 + \underset{\text{(Yellow ppt.)}}{PbCrO_4 \downarrow}$$

Copper (Cu^{2+})

Black ppt. of CuS dissolves in 50% nitric acid and a blue solution is obtained on addition of excess of NH_4OH.

$$3CuS + 8HNO_3 \longrightarrow 3Cu(NO_3)_2 + 4H_2O + 2NO + 3S$$

$$Cu(NO_3)_2 + 4NH_4OH \longrightarrow \underset{\text{(Blue solution)}}{[Cu(NH_3)_4]\,[NO_3]_2} + 4H_2O$$

1. *Potassium ferrocyanide test:*

$$[Cu(NH_3)_4]\,SO_4 + 4CH_3COOH \longrightarrow CuSO_4 + 4CH_3COONH_4$$

$$2CuSO_4 + K_4[Fe(CN)_6] \longrightarrow \underset{\text{(Chocolate brown colour)}}{Cu_2[Fe(CN)_6] \downarrow} + 2K_2SO_4$$

2. *Potassium iodide test:*

$$2CuSO_4 + 4KI \longrightarrow \underset{\text{(White ppt.)}}{Cu_2I_2 \downarrow} + \underset{\text{Brown colouration}}{I_2} + 2K_2SO_4$$

Arsenic (As^{3+})

The yellow residue of As_2S_3 is dissolved in conc. HNO_3 forming arsenic acid.

$$As_2S_3 + 10HNO_3 \longrightarrow \underset{\substack{\text{Arsenic acid}\\ \text{(Soluble)}}}{2H_3AsO_4} + 10NO_2 + 3S + 2H_2O$$

1. *Ammonium molybdate test:*

$$H_3AsO_4 + 12(NH_4)_2MoO_4 + 21HNO_3 \longrightarrow \underset{\substack{\text{Yellow ppt. of ammonium}\\ \text{arseno molybdate}}}{(NH_4)_3AsO_4.\,12MoO_3 \downarrow} + 21NH_4NO_3 + 12H_2O$$

2. *Magnesia mixture test:*

$$H_3AsO_4 + MgSO_4 + 3NH_4OH \longrightarrow \underset{\text{White ppt.}}{Mg(NH_4)_2AsO_4} + (NH_4)_2\,SO_4 + 3H_2O.$$

12.17. ANALYSIS OF GROUP III (IRON GROUP)

The cations present in this group are $\mathbf{Fe^{2+}}$, $\mathbf{Fe^{3+}}$, $\mathbf{Cr^{3+}}$ and $\mathbf{Al^{3+}}$. Only Fe^{2+}/Fe^{3+} and Al^{3+} are included in the syllabus of this class. These cations are precipitated as hydroxides by adding ammonium hydroxide in presence of ammonium chloride. Thus, **group reagent for this group is NH_4OH in the presence of NH_4Cl.**

PROCEDURE

In case, first and second groups are absent proceed for group III with the original solution. Take about 5 ml of the original solution and add 4–5 drops of conc. nitric acid. Boil the solution for sometime. Add to it about 2 g of solid NH_4Cl and boil again. Cool the solution under tap water. Add excess of ammonium hydroxide to it and shake. A ppt. shows the presence of some cation of group III. Filter the ppt. and wash with water. Note the colour of the ppt. If the ppt. is reddish brown in colour, it indicates the presence of Fe^{3+} and if the colour is white, it indicates the presence of Al^{3+}. Analyse the ppt. and draw inferences as in Table 12.14.

Table 12.14. Analysis of Group III (Fe^{3+} and Al^{3+})

Fe^{3+} (Reddish brown ppt.)	***Al^{3+} (White ppt.)***
Dissolve the reddish brown ppt. in dilute HCl, and divide the solution into two parts.	
Confirmation	**Confirmation**
1. **Potassium ferrocyanide test.** To one part of the above solution add potassium ferrocyanide solution. **Prussian blue colouration.**	1. **Lake test.** Dissolve the white ppt. in dilute hydrochloric acid. Add to it two drops of blue litmus solution. To this, add NH_4OH dropwise till blue colour develops. **Blue ppt. floating in the colourless solution.**
2. **Potassium sulphocyanide test.** To the second part, add a little potassium sulphocyanide solution. **Blood red colouration.**	2. **Cobalt nitrate test.** Perform charcoal cavity/ Cobalt nitrate test with the salt. **Blue mass.**

Notes:

1. **Test of Fe^{2+}.** *The addition of conc. nitric acid in the analysis of group III serves to oxidise Fe^{2+} ions to Fe^{3+} ions. Add conc. nitric acid only if the cation is Fe^{2+} otherwise the addition of nitric acid may be avoided. To test this, add a few drops of potassium ferricyanide solution to the original salt solution. A deep blue colouration shows Fe^{2+}.*
2. *Use sufficient quantity of ammonium chloride, otherwise the hydroxides of higher group may be precipitated along with the radicals of third group.*
3. *Add NH_4OH until the solution gives the smell of ammonia.*

Chemical Reactions Involved in the Analysis of Group III

The group III cations are precipitated as hydroxides on the addition of excess of ammonium hydroxide.

$$FeCl_3 + 3NH_4OH \longrightarrow 3NH_4Cl + \underset{\text{(Reddish brown ppt.)}}{Fe(OH)_3 \downarrow}$$

$$AlCl_3 + 3NH_4OH \longrightarrow 3NH_4Cl + \underset{\text{(White ppt.)}}{Al(OH)_3 \downarrow}$$

Iron (Fe^{3+})

The reddish brown ppt. of $Fe(OH)_3$ is dissolved in HCl.

$$Fe(OH)_3 + 3HCl \longrightarrow FeCl_3 + 3H_2O$$

1. *Potassium ferrocyanide test*:

$$4FeCl_3 + 3K_4[Fe(CN)_6] \longrightarrow 12KCl + \underset{\substack{\text{Ferric ferrocyanide} \\ \text{(Prussian blue)}}}{Fe_4[Fe(CN)_6]_3}$$

2. *Potassium sulphocyanide test*:

$$FeCl_3 + 3KCNS \longrightarrow 3KCl + \underset{\substack{\text{Ferric sulphocyanide} \\ \text{(Blood red colouration)}}}{Fe(CNS)_3}$$

Aluminium (Al^{3+})

1. *Lake test*:

$$Al(OH)_3 + 3HCl \longrightarrow AlCl_3 + 3H_2O \quad \text{...}\textit{dissolution}$$

$$AlCl_3 + 3NH_4OH \longrightarrow 3NH_4Cl + \underset{\substack{\text{Blue colour} \\ \text{adsorbs on} \\ \text{this ppt.}}}{Al(OH)_3 \downarrow}$$

12.18. ANALYSIS OF GROUP IV (ZINC GROUP)

The radicals present in this group are $\mathbf{Co^{2+}}$, $\mathbf{Ni^{2+}}$, $\mathbf{Mn^{2+}}$ and $\mathbf{Zn^{2+}}$. These are precipitated as sulphides by passing H_2S gas through the ammonical solution of the salt.

The group reagent for this group is H_2S gas in the presence of NH_4Cl and NH_4OH.

PROCEDURE

If there is no ppt. in the third group, then use the same ammonical solution for the fourth group. Pass H_2S gas through the solution. If some ppt. is formed, presence of some radical of group IV is indicated. Filter the ppt. and wash it with water. Note the colour of the ppt. and analyse the ppt. according to the Table 12.15.

Table 12.15. Analysis of Group IV Radicals (Co^{2+}, Ni^{2+}, Mn^{2+} and Zn^{2+})

Black ppt. (Co^{2+} or Ni^{2+}) *Observe the colour of the original salt. If the salt is purple or deep violet in colour perform confirmatory tests for Co^{2+} and if it is greenish perform confirmatory tests for Ni^{2+} with the original solution.*		***Buff (flesh) coloured ppt.*** ***Mn^{2+}***	***Dull white ppt.*** ***Zn^{2+}***
Confirmation of Co^{2+}	**Confirmation of Ni^{2+}**	**Confirmation of Mn^{2+}**	**Confirmation of Zn^{2+}**
1. Potassium nitrite test To one part of the O.S. add ammonium hydroxide to neutralise the solution. Add acetic acid and a crystal of potassium nitrite. Warm. **A yellow ppt.** is formed.	**1. Dimethyl glyoxime test** To one part of O.S. add amm. hydroxide soln. and few drops of dimethyl glyoxime. **Bright rose red ppt.** is obtained.	**1. Sodium hydroxide-Br_2 test** To the O.S. add NaOH soln. Shake. A **white ppt.** is formed. Add **Br_2** water to *white ppt.* **It turns black or brown.**	**1. Sodium hydroxide test** To one part of O.S. add sodium hydroxide solution dropwise. A **white ppt.** is formed. Add more of NaOH. The white ppt. dissolves.
2. Amm. thiocyanate ether test To another part add ether (1 ml). Add a crystal of ammonium thiocyanate, shake. Allow to settle. **Blue colour** in ethereal layer confirms Co^{2+}.	**2. Sodium hydroxide-Br_2 test** To another part add sodium hydroxide (in excess) and bromine water. Boil. **A black ppt.** is formed.	**2. Lead peroxide test** To black ppt. obtained in above test add conc. HNO_3 and lead peroxide. Boil, cool and allow to settle. **Pink-coloured** soln. is formed.	**2. Pot. ferrocyanide test** To another part, add pot. ferrocyanide soln. **White or bluish white ppt.** is formed.
3. Borax bead test Perform borax bead test with the salt. A **blue bead** is formed.	**3. Borax bead test** Perform borax bead test with the salt. **Brown bead** in oxidizing and **grey bead** in reducing flame is obtained.	**3. Borax bead test** Perform borax bead test with the salt. **Pinkish bead** in oxidizing flame and **colourless bead** in reducing flame.	**3. Charcoal Cavity/ Cobalt Nitrite Test** Perform Charcoal Cavity/Cobalt Nitrate test with the salt. **Greenish residue** is obtained.

Chemical Reactions Involved in the Analysis of Group IV

Passing of H_2S gas through the group III solution will precipitate the radicals Co^{2+}, Ni^{2+}, Mn^{2+} and Zn^{2+} as their sulphides. Formation of black ppt. (CoS or NiS) indicates cobalt or nickel. Formation of buff-coloured ppt. (MnS) indicates manganese and dirty white ppt. (ZnS) indicates zinc.

$$Co(OH)_2 + H_2S \longrightarrow 2H_2O + CoS\downarrow \; \textit{(Black ppt.)}$$

$$Ni(OH)_2 + H_2S \longrightarrow 2H_2O + NiS\downarrow \; \textit{(Black ppt.)}$$

$$Zn(OH)_2 + H_2S \longrightarrow 2H_2O + ZnS\downarrow \; \textit{(White ppt.)}$$

$$Mn(OH)_2 + H_2S \longrightarrow 2H_2O + MnS\downarrow \; \textit{(Buff-coloured ppt.)}$$

Cobalt (Co^{2+})

1. *Potassium nitrite test:*

$$\underset{\text{(O.S.)}}{CoCl_2} + 2KNO_2 \longrightarrow 2KCl + \underset{\text{Cobaltous nitrite}}{Co(NO_2)_2}$$

$$KNO_2 + CH_3COOH \longrightarrow CH_3COOK + HNO_2$$

$$Co(NO_2)_2 + 2HNO_2 \longrightarrow \underset{\text{Cobalt nitrite}}{Co(NO_2)_3} + H_2O + NO$$

$$Co(NO_2)_3 + 3KNO_2 \longrightarrow \underset{\substack{\text{Pot. cobalti nitrite}\\ \text{(Yellow ppt.)}}}{K_3[Co(NO_2)_6]}$$

2. *Ammonium thiocyanate ether test:*

On addition of ether and a crystal of ammonium thiocyanate (shaking and allowing to stand), a blue colour is obtained in the ethereal layer.

$$CoCl_2 + 4NH_4CNS \longrightarrow \underset{\text{(Blue-colouration)}}{(NH_4)_2[Co(CNS)_4]} + 2NH_4Cl$$

Nickel (Ni^{2+})

1. *Dimethyl glyoxime test (with O.S.):*

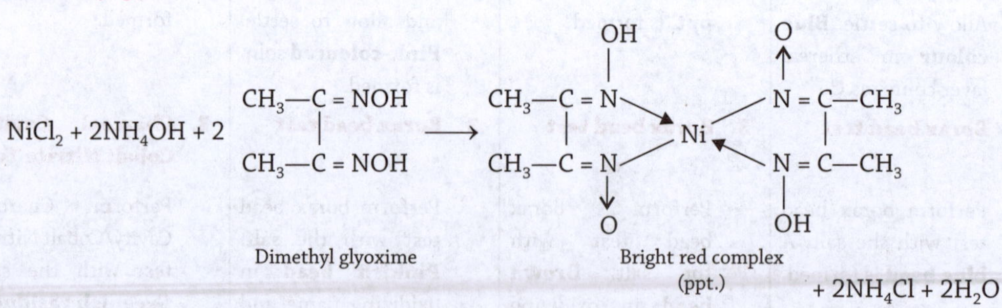

2. *Sodium hydroxide-bromine water test*

$$NiCl_2 + 2NaOH \longrightarrow 2NaCl + \underset{\text{(green ppt.)}}{Ni(OH)_2\downarrow}$$

$$Br_2 + H_2O \longrightarrow 2HBr + [O]$$

$$2Ni(OH)_2 + H_2O + [O] \longrightarrow \underset{\text{Nickelic hydroxide (Black ppt.)}}{2Ni(OH)_3 \downarrow}$$

Manganese (Mn^{2+})

Manganese sulphides dissolves in dil. HCl forming manganese chloride, and H_2S is boiled off.

$$MnS + 2HCl \longrightarrow MnCl_2 + H_2S \uparrow$$

1. *NaOH and Br_2 water test*

$$MnCl_2 + 2NaOH \longrightarrow \underset{\text{White ppt.}}{Mn(OH)_2 \downarrow} + 2NaCl$$

The white ppt. of manganese hydroxide turns brown on adding Br_2 water due to its oxidation to brown manganic hydroxide, $MnO(OH)_2$

$$Br_2 + H_2O \longrightarrow 2HBr + [O]$$

$$Mn(OH)_2 + [O] \longrightarrow \underset{\text{Brown ppt.}}{MnO(OH)_2 \downarrow}$$

2. *PbO_2 test*

$$MnS + 2HNO_3 \longrightarrow Mn(NO_3)_2 + H_2S$$

$$2Mn(NO_3)_2 + 5PbO_2 + 6HNO_3 \longrightarrow \underset{\text{Pink solution}}{2HMnO_4} + 5Pb(NO_3)_2 + 2H_2O$$

Zinc (Zn^{2+})

The precipitate of ZnS obtained in Group IV is white.

The white ppt. of ZnS dissolves in dil. HCl, and H_2S is boiled off.

$$ZnS + 2HCl \longrightarrow ZnCl_2 + H_2S \uparrow$$

1. *NaOH test*

$$ZnCl_2 + 2NaOH \longrightarrow \underset{\text{White ppt.}}{Zn(OH)_2 \downarrow} + 2NaCl$$

$$\underset{\text{White ppt.}}{Zn(OH)_2} + 2NaOH \longrightarrow \underset{\text{Soluble}}{Na_2ZnO_2} + 2H_2O$$

2. *Potassium ferrocyanide test*

$$2ZnCl_2 + K_4[Fe(CN)_6] \longrightarrow \underset{\text{White or Bluish-white ppt.}}{Zn_2[Fe(CN)_6] \downarrow} + 4KCl$$

12.19. ANALYSIS OF GROUP V (CALCIUM GROUP)

Group V consists of three radicals: **Ba^{2+}, Sr^{2+}** and **Ca^{2+}**. These cations are precipitated as their carbonates.

Group reagent for this group is $(NH_4)_2CO_3$ in the presence of NH_4Cl and NH_4OH.

PROCEDURE

If the fourth group is absent, then proceed for radicals of group V.

To the O.S. add 2–3 gm of solid NH_4Cl, boil, cool and add NH_4OH till the solution smells of ammonia. Then add $(NH_4)_2CO_3$ solution. Appearance of white ppt. indicates the presence of group V cation. Filter and wash the ppt. with water. Dissolve the ppt. in hot dil. acetic acid. Divide the solution into three parts and proceed as in Table 12.16.

Table 12.16. Analysis of Group V (Ba^{2+}, Sr^{2+}, Ca^{2+})

Ba^{2+}	*Sr^{2+}*	*Ca^{2+}*
	Test for Sr^{2+} only if Ba^{2+} is absent.	**Test for Ca^{2+} only if Ba^{2+} and Sr^{2+} are absent.**
1. Potassium chromate test	**1. Amm. sulphate test**	**1. Amm. oxalate test**
To one part of the solution, add a few drops of potassium chromate solution. **Yellow ppt.**	To the second part of the solution, add 1 ml of amm. sulphate solution and warm. **White ppt.**	To the third portion of the solution, add 1–2 ml of amm. oxalate solution. Add a little amm. hydroxide to it and scratch the sides. **White ppt.**
2. Flame test	**2. Flame test**	**2. Flame test**
Perform the flame test with the original salt. **Grassy green flame.**	Perform the flame test with the original salt. **Crimson red flame.**	Perform the flame test with the original salt. **Brick red flame.**

Notes:

1. *Proceed to test for group V cations in the order, Ba^{2+}, Sr^{2+}, Ca^{2+}. If Ba^{2+} is confirmed, do not test for Sr^{2+} or Ca^{2+}. Similarly if Sr^{2+} is confirmed, do not test for Ca^{2+}.*
2. *Original solution can be preferably used for testing Sr^{2+} and Ca^2*

Chemical Reactions Involved in the Analysis of Group V Radicals

When $(NH_4)_2CO_3$ is added to a salt solution containing NH_4Cl and NH_4OH, the carbonates of Ba^{2+}, Sr^{2+} and Ca^{2+} are precipitated.

$$BaCl_2 + (NH_4)_2CO_3 \longrightarrow BaCO_3 \downarrow + 2NH_4Cl$$

$$SrCl_2 + (NH_4)_2CO_3 \longrightarrow SrCO_3 \downarrow + 2NH_4Cl$$

$$CaCl_2 + (NH_4)_2CO_3 \longrightarrow CaCO_3 \downarrow + 2NH_4Cl$$

The insoluble carbonates dissolve in acetic acid due to formation of corresponding soluble acetates.

Barium (Ba^{2+})

White ppt. of $BaCO_3$ dissolves in hot dilute acetic acid.

$$BaCO_3 + 2CH_3COOH \longrightarrow (CH_3COO)_2Ba + CO_2 \downarrow + H_2O$$

1. *Potassium chromate test*

$$(CH_3COO)_2Ba + K_2CrO_4 \longrightarrow 2CH_3COOK + \underset{\text{(yellow ppt.)}}{BaCrO_4 \downarrow}$$

2. *Flame test*

Barium imparts grassy green colour to the flame.

Strontium (Sr^{2+})

White ppt. of $SrCO_3$ dissolves in hot dilute acetic acid.

$$SrCO_3 + 2CH_3COOH \longrightarrow (CH_3COO)_2Sr + CO_2\uparrow + H_2O$$

1. *Ammonium sulphate test*

$$(CH_3COO)_2Sr + (NH_4)_2SO_4 \longrightarrow 2CH_3COONH_4 + \underset{\text{(white ppt.)}}{SrSO_4\downarrow}$$

2. *Flame test*

Strontium produces crimson red flame

Calcium (Ca^{2+})

White ppt. of $CaCO_3$ dissolves in hot dil. acetic acid.

$$CaCO_3 + 2CH_3COOH \longrightarrow (CH_3COO)_2Ca + CO_2\uparrow + H_2O$$

1. *Ammonium oxalate test*

$$(CH_3COO)_2\,Ca + (NH_4)_2C_2O_4 \longrightarrow 2CH_3COONH_4 + \underset{\text{(white ppt.)}}{CaC_2O_4\downarrow}$$

2. *Flame test*

Calcium imparts *brick red colour* to the flame.

12.20. ANALYSIS OF GROUP VI (MAGNESIUM GROUP)

Table 12.17. Analysis of Group VI (Mg^{2+})

Mg^{2+}
1. Ammonium phosphate test To a part of the original solution add some solid NH_4Cl and NH_4OH in slight excess. Then add ammonium phosphate solution and scratch the sides of the test tube with a glass rod. **A white ppt. confirms Mg^{2+}.**
2. Charcoal cavity cobalt nitrate test. Perform charcoal cavity cobalt nitrate test with the original salt. A **pink mass** *is obtained.*

Chemical Reactions Involved in Confirmation of Mg^{2+}

1. *Ammonium phosphate test*

$$MgCl_2 + NH_4OH + (NH_4)_2HPO_4 \longrightarrow \underset{\text{(White ppt.)}}{Mg(NH_4)PO_4\downarrow} + 2NH_4Cl + H_2O$$

SPECIMEN RECORD OF ANALYSIS OF A SALT

EXPERIMENT 12.1

AIM: ***To analyse the given salt for acidic and basic radicals.***

Experiment	*Observations*	*Inference*
1. Physical examination :		
(*a*) Noted the colour of the given salt.	White	Cu^{2+}, Fe^{2+}, Fe^{3+}, Co^{2+}, Ni^{2+}, Mn^{2+} absent.
(*b*) Noted the smell of the salt.	No specific odour	NH_4^+, S^{2-} and CH_3COO^- may be absent.
2. Dry heating test Heated a pinch of the salt in a dry test tube and noted the following observations :		
(a) Gas evolved	A colourless gas having pungent smell evolved. The gas gave white fumes when a glass rod dipped in aqueous ammonia is brought near the mouth of the test tube.	Some hydrated Cl^-, Br^- or I^- may be present.
(b) Sublimation	No sublimate formed.	Ammonium halides, aluminium chloride, iodide may be absent.
(c) Decrepitation	No crackling sound observed.	Lead nitrate, barium nitrate, sodium chloride, potassium chloride and potassium iodide may be absent.
(d) Fusion	Salt does not fuse.	Alkali (sodium, potassium) salts may be absent.
(e) Colour of the residue	White	Zn^{2+}, Pb^{2+} may be absent.
3. Charcoal cavity test Mixed a pinch of the salt with double the quantity of Na_2CO_3 and heated the mixture on a charcoal cavity in the reducing flame.	White residue.	Zn^{2+}, Pb^{2+}, Mn^{2+} etc. may be absent.
4. Cobalt nitrate test To the above white residue added a drop of cobalt nitrate solution and heated in oxidising flame.	No characteristic colour.	Zn^{2+}, Mg^{2+}, Al^{3+}, PO_4^{3-}, may be absent.

Experiment	*Observations*	*Inference*
5. Flame test Prepared a paste of the salt in conc. HCl and performed flame test.	Persistent grassy green flame on prolonged heating.	Ba^{2+} present.
6. Borax bead test Did not perform this test since the given salt was white.	—	Cu^{2+}, Ni^{2+}, Fe^{3+}, Mn^{2+} may be absent.
7. Dil. sulphuric acid test Treated a pinch of the salt with dil. H_2SO_4 and warmed.	No gas evolved.	CO_3^{2-}, S^{2-}, NO_2^-, SO_3^{2-} may be absent.
8. $KMnO_4$ test To a pinch of the salt added dil. H_2SO_4 warm and then a drop of $KMnO_4$ solution.	Pink colour of $KMnO_4$ was discharged with evolution of yellowish green gas having characteristic pungent smell.	Cl^- may be present.
9. Conc. sulphuric acid test Heated a pinch of the salt with conc. sulphuric acid.	A colourless gas having pungent smell evolved. The gas gave white fumes when a glass rod dipped in NH_4OH solution was brought near the mouth of the test tube.	Cl^- may be present.
10. Confirmatory tests for chloride *Preparation of Sodium Carbonate extract* Took about 1 g of the salt, added double the quantity of solid Na_2CO_3 and about 10 ml of water. Boiled for some time, cooled and filtered.	Labelled the filtrate obtained as sodium carbonate extract (S.E)	
(a) Silver nitrate test Acidified a portion of S.E with HNO_3. Boiled, cooled and added silver nitrate solution.	A white precipitate formed which was soluble in NH_4OH.	Cl^- confirmed.
(b) Manganese dioxide test Heated a small quantity of the salt with solid manganese dioxide and conc. H_2SO_4.	A greenish yellow gas having pungent smell evolved. The gas turned starch-iodide paper blue.	Cl^- confirmed.
11. Heated a pinch of salt with conc. NaOH solution	No ammonia gas evolved.	NH_4^+ absent.
12. Preparation of Original Solution (O.S.) Shook a pinch of the salt with water.	Solution obtained	Labelled it as Original Solution (O.S.)

Experiment	*Observations*	*Inference*
13. To a part of the O.S. added 1–2 mls of dilute hydrochloric acid.	No ppt. formed.	Group I absent. (Pb^{2+} absent)
14. Through a part of the above solution, passed H_2S gas.	No ppt. formed.	Group II absent (Pb^{2+}, Cu^{2+} absent)
15. To the remaining solution, added a pinch of solid ammonium chloride. Boiled the solution, cooled it and added excess of ammonium hydroxide solution.	No ppt. formed.	Group III absent. (Fe^{3+}, Al^{3+} absent)
16. Through a part of this solution, passed H_2S gas.	No ppt. formed.	Group IV absent. (Zn^{2+}, Mn^{2+}, Ni^{2+}, absent)
17. To the remaining ammonical solution added ammonium carbonate solution.	White ppt. formed.	Group V present. (Ca^{2+}, Ba^{2+}, Sr^{2+} may be present)
18. Confirmatory test for Barium Filtered the above white ppt. Dissolved the ppt. in hot dilute acetic acid.		
(*a*) ***Pot. chromate test.*** To one part of the above solution, added a few drops of pot. chromate solution.	Yellow ppt.	Ba^{2+} confirmed.
(*b*) ***Flame test.*** Performed flame test with the salt.	Persistent grassy green flame on prolonged heating.	Ba^{2+} confirmed.

RESULT

Acid radical: Cl^-

Basic radical: Ba^{2+}.

EXPERIMENT 12.2

AIM: ***To analyse the given salt for acidic and basic radicals.***

Experiment	*Observations*	*Inference*
1. Physical examination		
(*a*) Noted the colour of the given salt.	White	Cu^{2+}, Fe^{3+}, Co^{2+}, Ni^{2+}, Mn^{2+}, absent.
(*a*) Noted the smell of the salt.	No specific odour	NH_4^+, S^{2-} and CH_3COO^- may be absent.
2. Dry heating test Heated a pinch of the salt in a dry test tube and noted the following :		

Experiment	*Observations*	*Inference*
(a) Gas evolved	A colourless, odourless gas evolved which turned lime water milky.	CO_3^{2-} may be present.
(b) Sublimation	No sublimate formed.	Ammonium halides, iodide may be absent.
(c) Decrepitation	No crackling sound observed.	Lead nitrate, barium nitrate, sodium chloride, potassium chloride and potassium iodide may be absent.
(d) Colour of the residue	Yellow when hot and white when cold.	Zn^{2+} may be present.
3. Charcoal cavity test Mixed a pinch of the salt with double the quantity of Na_2CO_3 and heated the mixture on a charcoal cavity in the reducing flame.	Yellow when hot and white when cold.	Zn^{2+} may be present.
4. Cobalt nitrate test To the above white residue added a drop of cobalt nitrate solution and heated in oxidizing flame.	Green residue.	Zn^{2+} may be present.
5. Flame test Prepared a paste of the salt in conc. HCl and performed flame test.	Green flashes seen with naked eye.	Zn^{2+} may be present.
6. Borax bead test Did not perform this test since the given salt was white.	—	Cu^{2+}, Ni^{2+}, Fe^{2+}, Fe^{3+}, Mn^{2+} may be absent.
7. Dil. Sulphuric acid test Treated a pinch of the salt with dil. H_2SO_4 and warmed.	Colourless, odourless gas evolved with brisk effervescence, turned lime water milky.	CO_3^{2-} present
Shook a pinch of salt with water taken in test tube.	Salt did not dissolve.	Insoluble CO_3^{2-} indicated.
8. $KMnO_4$ test To a pinch of the salt added dilute H_2SO_4 warm and then a drop of $KMnO_4$ solution.	Pink colour of $KMnO_4$ was not discharged.	Cl^-, Br^-, I^-, Fe^{2+}, $C_2O_4^{2-}$ are absent.
9. Conc. Sulphuric acid test Did not perform this test because the salt reacted with dil. H_2SO_4.	—	Cl^-, Br^-, I^-, NO_3^-, CH_3COO^-, $C_2O_4^{2-}$ are absent.
10. Confirmatory tests for carbonate		
(a) Shook a pinch of the salt with water.	Salt did not dissolve.	Insoluble carbonate indicated.

Experiment	Observations	Inference
(b) To the salt added dil. HCl.	Brisk effervescence with evolution of colourless, odourless gas which turned lime water milky.	Insoluble carbonate confirmed.
11. Heated a pinch of salt with conc. NaOH solution	No ammonia gas evolved.	NH_4^+ absent.
12. Preparation of Original solution (O.S.)		
(a) Shook a pinch of the salt with water.	Insoluble	Labelled it as O.S.
(b) Shook a pinch of the salt in dil. HCl.	Clear solution obtained.	
13. As the O.S. is prepared in dil. HCl.		Group I absent. (Pb^{2+} absent)
14. Through a part of O.S. passed H_2S gas.	No ppt. formed.	Group II absent (Pb^{2+}, Cu^{2+} absent).
15. To the remaining solution, added a pinch of solid ammonium chloride. Boiled the solution, cooled it and added excess of ammonium hydroxide solution.	No ppt. formed	Group III absent. (Fe^{3+}, Al^{3+} absent).
16. Through a part of this solution, passed H_2S gas.	Dull white ppt. formed.	Group IV present. (Zn^{2+} present)
17. Confirmatory tests for Zn^{2+} ion Dissolved the above dull white ppt. in dil HCl. Boiled off H_2S. Divided the solution into two parts.		
(a) To one part added NaOH solution dropwise.	White ppt. soluble in excess of NaOH.	Zn^{2+} confirmed.
(b) To another part, added potassium ferrocyanide solution.	Bluish white ppt.	Zn^{2+} confirmed.

RESULT

Acid Radical : CO_3^{2-}

Basic Radical : Zn^{2+}.

Table 12.18. List of Common White Salts

Name of the Salt	*Basic Radical*	*Acidic Radical*
Lead Nitrate	Pb^{2+}	NO_3^-
Lead Acetate	Pb^{2+}	CH_3COO^-
Zinc Carbonate	Zn^{2+}	CO_3^{2-}
Zinc Sulphide	Zn^{2+}	S^{2-}
Zinc Nitrate	Zn^{2+}	NO_3^-
Zinc Acetate	Zn^{2+}	CH_3COO^-
Zinc Chloride	Zn^{2+}	Cl^-
Zinc Bromide	Zn^{2+}	Br^-
Zinc Sulphate	Zn^{2+}	SO_4^{2-}
Calcium Sulphite	Ca^{2+}	SO_3^{2-}
Calcium Carbonate	Ca^{2+}	CO_3^{2-}
Calcium Chloride	Ca^{2+}	Cl^-
Calcium Bromide	Ca^{2+}	Br^-
Calcium Acetate	Ca^{2+}	CH_3COO^-
Calcium Nitrate	Ca^{2+}	NO_3^-
Barium Carbonate	Ba^{2+}	CO_3^{2-}
Barium Chloride	Ba^{2+}	Cl^-
Barium Nitrate	Ba^{2+}	NO_3^-
Strontium Carbonate	Sr^{2+}	CO_3^{2-}
Strontium Chloride	Sr^{2+}	Cl^-
Strontium Nitrate	Sr^{2+}	NO_3^-
Magnesium Carbonate	Mg^{2+}	CO_3^{2-}
Magnesium Acetate	Mg^{2+}	CH_3COO^-
Magnesium Sulphate	Mg^{2+}	SO_4^{2-}
Ammonium Carbonate	NH_4^+	CO_3^{2-}
Ammonium Chloride	NH_4^+	Cl^-
Ammonium Bromide	NH_4^+	Br^-
Ammonium Iodide	NH_4^+	I^-
Ammonium Nitrate	NH_4^+	NO_3^-
Ammonium Sulphate	NH_4^+	SO_4^{2-}
Ammonium Phosphate	NH_4^+	PO_4^{3-}

Table 12.19. List of Common Coloured Salts

Name of the Salt	*Basic Radical*	*Acidic Radical*
Copper Sulphate	Cu^{2+}	SO_4^{2-}
Ferrous Sulphate	Fe^{2+}	SO_4^{2-}
Manganese Chloride	Mn^{2+}	Cl^-
Cobalt Nitrate	Co^{2+}	NO_3^-
Nickel Carbonate	Ni^{2+}	CO_3^{2-}
Copper Acetate	Cu^{2+}	CH_3COO^-
Cobalt Acetate	Co^{2+}	CH_3COO^-
Manganese Sulphate	Mn^{2+}	SO_4^{2-}
Cobalt Sulphate	Co^{2+}	SO_3^{2-}
Copper Chloride	Cu^{2+}	Cl^-
Nickel Sulphate	Ni^{2+}	SO_4^{2-}
Copper Carbonate	Cu^{2+}	CO_3^{2-}

VIVA QUESTIONS WITH ANSWERS

Q. 1. What is qualitative analysis?

Ans. The type of analysis that deals with the methods which are used to determine the constituents of a compound.

Q. 2. What is a radical?

Ans. A radical may be defined as an atom or group of atoms which carries charge and behaves as a single unit in chemical reactions.

Q. 3. What are acidic and basic radicals?

Ans. Radicals carrying positive charge are called basic radicals and those carrying negative charge are called acidic radicals.

Q. 4. What type of bond is generally present in an inorganic salt?

Ans. Electrovalent bond.

Q. 5. Why do inorganic salt ionise when dissolved in water?

Ans. Due to the high dielectric constant of water, the force of attraction holding the two ions in a salt decreases. Thus, the two ions separate. The ions are further stabilized by hydration.

Q. 6. Name the coloured basic radicals.

Ans. Cu^{2+}, Fe^{2+}, Fe^{3+}, Cr^{3+}, Ni^{2+}, Co^{2+} and Mn^{2+}.

Q. 7. What is the colour of iron salts?

Ans. Ferrous salts are usually light green while ferric salts are generally brown.

Q. 8. Name any iron salt which is light green.

Ans. Ferrous sulphate.

Q. 9. What is the colour of nickel salts?

Ans. Bluish green or green.

Q. 10. What is the colour of manganese salts?

Ans. Light pink or flesh colour.

Q. 11. Name the basic radicals which are absent, if the given salt is white.

Ans. Cu^{2+}, Fe^{2+}, Fe^{3+}, Cr^{3+}, Ni^{2+}, Co^{2+} and Mn^{2+}.

Q. 12. Why does a salt containing lead turn black in colour, when placed for a long time in laboratory ?

Ans. Due to the formation of black lead-sulphide by the action of H_2S in the atmosphere.

Q. 13. Name some salts which produce crackling sound when heated.

Ans. Lead nitrate, barium nitrate, potassium bromide, sodium chloride.

Q. 14. What is sublimation?

Ans. It is the process by which a salt directly changes into gaseous phase without melting, when heated. On cooling vapours condense back to the solid state.

Q. 15. Tell the importance of preliminary tests in qualitative analysis.

Ans. Sometimes, preliminary tests give authentic information about an ion in the salt. For example, crimson red colour in flame test shows the presence of strontium. In a charcoal cavity test, brown residue shows the presence of cadmium in a salt and so on.

Q. 16. How is dry heating test performed and what information you get if the residue changes to yellow when hot?

Ans. In dry heating test, the salt is heated in a dry test tube. Yellow residue when hot shows the presence of zinc.

Q. 17. What is the expected information when copper sulphate is heated in a dry test tube?

Ans. A white residue is formed and water condenses on the colder walls of the test tube.

Q. 18. Name the radical which produces CO_2 on heating.

Ans. Carbonate.

Q. 19. What is the colour of residue when zinc salt is heated?

Ans. A residue yellow when hot and white when cold is formed.

Q. 20. If the residue in dry heating test is white, name the radicals which are absent.

Ans. Cu^{2+}, Fe^{2+}, Ni^{2+}, Mn^{2+}, Co^{2+}, Cr^{3+}, Cd^{2+}, Zn^{2+} and Pb^{2+}.

Q. 21. How is charcoal cavity test performed? Describe the chemistry for the formation of incrustation as well as metallic bead.

Ans. The salt is mixed with the double the quantity of sodium carbonate and the mixture is heated in the charocal cavity in luminous flame (reducing flame).

$$Pb(NO_3)_2 + Na_2CO_3 \longrightarrow PbCO_3 + 2NaNO_3$$

$$PbCO_3 \longrightarrow CO_2 + \underset{\text{(incrustation)}}{\underset{\text{Brown-hot}}{PbO}}$$

$$PbO + C \longrightarrow \underset{\text{(Bead)}}{Pb} + CO\uparrow$$

Q. 22. Which flame is used in charcoal cavity test? How is it obtained?

Ans. A reducing flame is used in charcoal cavity test. It is obtained by closing the air holes of the Bunsen burner.

Q. 23. Why should we avoid excess of cobalt nitrate in cobalt nitrate test?

Ans. Excess of cobalt nitrate is avoided because it forms black cobalt oxide in the oxidising flame. This colour masks the other colours which might be produced during the test.

Q. 24. In the flame test, sodium imparts yellow colour to the flame while magnesium does not impart any colour. Why?

Ans. In case of magnesium, the energy of flame is unable to promote the electron to higher energy level, hence, no colour is imparted to the flame.

Q. 25. What is the chemistry of the flame test?

Ans. In flame test, the valence electron of the atom gets excited and jumps to the higher level. When the electron jumps back to the ground state, the radiation is emitted whose frequency falls in the visible region.

Q. 26. What is the function of blue glass in flame test?

Ans. The blue glass can absorb a part or whole of the coloured light in certain cases. Therefore, the flame appears to be of different colour when viewed through blue glass. This helps in identification of some basic radicals.

Q. 27. Why do we use conc. HCl in preparing a paste of the salt for the flame test?

Ans. In order to convert metal salts into metal chlorides which are more volatile than other salts.

Q. 28. Why can't we use glass rod instead of platinum wire for performing flame test?

Ans. This is because glass contains sodium silicate which imparts its own golden yellow colour to the flame.

Q. 29. Why is platinum metal preferred to other metals for flame test?

Ans. Because platinum does not react with acids and does not itself impart any characteristic colour to the flame.

Q. 30. Why do barium salts not impart colour to the flame immediately?

Ans. Because barium chloride is less volatile, it imparts colour to the flame after some time.

Q. 31. Why should we avoid the use of platinum wire for testing lead salts?

Ans. Because lead combines with platinum and the wire gets corroded.

Q. 32. Why should only a particle or two of the given salt should be touched with the bead in borax bead test?

Ans. If salt is used in excess an opaque bead is formed.

Q. 33. Why borax bead test is not applicable in case of white salts?

Ans. White salts do not form coloured meta-borates.

Q. 34. What is Nessler's Reagent?

Ans. It is a solution of mercuric iodide in potassium iodide. Its formula is $K_2[HgI_4]$.

Q. 35. Name the acid radicals detected with dil. H_2SO_4.

Ans. CO_3^{2-}, S^{2-}, SO_3^{2-}, NO_2^-.

Q. 36. Why is dil. H_2SO_4 preferred while testing acid radicals over dil. HCl?

Ans. When the salt is treated with HCl, during reaction HCl gas is also given out along with the gas evolved by the salt. So the actual gas cannot be identified whereas with H_2SO_4, no such problem arises.

Q. 37. Which anions are detected by conc. H_2SO_4 test?

Ans. Cl^-, Br^-, I^-, NO_3^-, CH_3COO^-, $C_2O_4^{2-}$.

Q. 38. Name the radicals which are always tested using water extract.

Ans. NO_3^-, NO_2^- and CH_3COO^-.

Q. 39. How is sodium carbonate extract prepared?

Ans. The salt is mixed with double the amount of solid Na_2CO_3 and about 20 ml of distilled water. It is then boiled till it is reduced to one-third, and then filtered. The filtrate is sodium carbonate extract or (S.E.).

Q. 40. CO_2 and SO_2 both turn lime water milky. How will you distinguish between them?

Ans. By passing through acidified $K_2Cr_2O_7$ solution. SO_2 turns $K_2Cr_2O_7$ green while CO_2 has no effect.

Q. 41. NO_2 and Br_2 both are brown in colour. How will you distinguish between them?

Ans. By passing through $FeSO_4$ solution. NO_2 turns $FeSO_4$ soln. black while Br_2 has no effect.

Q. 42. How will you test the presence of carbonate?

Ans. Treat a small quantity of the mixture with dil. H_2SO_4. CO_2 gas is evolved. When the gas is passed through lime water, it is turned milky.

$$Na_2CO_3 + 2HCl \longrightarrow 2NaCl + H_2O + CO_2.$$

Q. 43. What is lime water?

Ans. A solution of $Ca(OH)_2$ in water is called lime water.

Q. 44. What will happen if excess of CO_2 is passed through lime water?

Ans. The white ppt. of $CaCO_3$ changes into soluble calcium bicarbonate and the milkiness, therefore, disappears.

$$CaCO_3 + CO_2 + H_2O \longrightarrow Ca(HCO_3)_2.$$

Q. 45. How do you test for sulphide?

Ans. Warm the salt with dil. H_2SO_4. H_2S gas is evolved. It turns a paper dipped in lead acetate black.

$$Na_2S + 2HCl \longrightarrow 2NaCl + H_2S$$

$$Pb(CH_3COO)_2 + H_2S \longrightarrow PbS + 2CH_3COOH.$$

Q. 46. Name a gas other than CO_2 which turns lime water milky.

Ans. Sulphur dioxide gas (SO_2).

Q. 47. All nitrates on heating with conc. H_2SO_4 in presence of paper pallet evolve NO_2 gas. What is the function of paper pallet?

Ans. Paper pallet (carbon) reduces HNO_3 to NO_2

$$KNO_3 + H_2SO_4 \longrightarrow KHSO_4 + HNO_3$$

$$4HNO_3 + C \longrightarrow 2H_2O + 4NO_2 + CO_2.$$

Q. 48. How is ring test performed for nitrates?

Ans. To the salt solution, freshly prepared ferrous sulphate solution is added and then sulphuric acid (conc.) is added along the walls of the tube. A dark brown ring is formed at the junction of the two solutions.

Q. 49. Why is the hot reaction mixture in case of conc. H_2SO_4 test not thrown into the sink?

Ans. In order to avoid spurting, due to which H_2SO_4 may fly and spoil clothes and may result into serious injuries.

Q. 50. What is Tollen's reagent?

Ans. Ammonical $AgNO_3$ solution is called Tollen's reagent.

Q. 51. Why does a dark brown ring form at the junction of two layers in ring test for nitrates?

Ans. H_2SO_4 being heavier forms the lower layer and reacts only with a small amount of nitrate and $FeSO_4$ at its surface, therefore, a brown ring appears at the junction of the two layers.

Q. 52. What is the formula of sodium nitroprusside?

Ans. $Na_2[Fe(CN)_5NO]$.

Q. 53. What is chromyl chloride test?

Ans. Heat a small amount of the mixture with conc. H_2SO_4 and solid $K_2Cr_2O_7$ in a dry test tube. Deep brownish red vapours of chromyl chloride are formed. Pass these vapours in water. A yellow sol. of H_2CrO_4 is formed. Add to this solution NaOH, acetic acid and lead acetate, a yellow ppt. confirms chloride in the mixture.

Q. 54. What is the chemistry of carbon disulphide test for a bromide or iodide?

Ans. To a part of the soda extract add dil. HCl. Now to this add small amount of CS_2 and excess of chlorine water and shake the solution well. Chlorine displaces bromine or iodine from the bromide or iodide, which dissolves in carbon disulphide to produce orange or violet colouration.

$$2KBr + Cl_2 \longrightarrow 2KCl + Br_2$$
$$2KI + Cl_2 \longrightarrow 2KCl + I_2.$$

Q. 55. Why do bromides and iodides not respond to chromyl chloride test?

Ans. Because chromyl bromide (CrO_2Br_2) and chromyl iodide (CrO_2I_2) compounds are not formed, instead of these bromine and iodine are evolved.

$$K_2Cr_2O_7 + 6KI + 7H_2SO_4 \longrightarrow 3I_2 + Cr_2(SO_4)_3 + 4K_2SO_4 + 7H_2O$$
$$K_2Cr_2O_7 + 6KBr + 7H_2SO_4 \longrightarrow 3Br_2 + Cr_2(SO_4)_3 + 4K_2SO_4 + 7H_2O.$$

Q. 56. Describe the chemistry of match stick test.

Ans. In match stick test, the sulphate is reduced to sulphide by carbon of match stick which then gives violet colour with sodium nitroprusside solution.

$$ZnSO_4 + Na_2CO_3 \longrightarrow ZnCO_3 + Na_2SO_4$$
$$Na_2SO_4 + \underset{\text{(Match stick)}}{4C} \longrightarrow Na_2S + 4CO$$
$$Na_2S + Na_2[Fe(CN)_5NO] \longrightarrow \underset{\text{Purple colour}}{Na_4[Fe(CN)_5NOS]}$$

Q. 57. Why does iodine give a blue colour with starch solution?

Ans. The blue colour is due to the formation of a complex between iodine and starch.

Q. 58. Why is original not prepared in conc. HNO_3?

Ans. HNO_3 is an oxidising agent which on decomposition gives oxygen. A yellow ppt. of sulphur is obtained in presence of HNO_3 when H_2S is passed.

$$H_2S + 2HNO_3 \longrightarrow 2NO_2 + 2H_2O + S.$$

Q. 59. Name group reagents for different groups.

Ans. Group I—Dil. HCl.

Group II—H_2S in the presence of dil. HCl.

Group III—NH_4OH in the presence of NH_4Cl.

Group IV—H_2S in the presence of NH_4OH.

Group V—$(NH_4)_2\,CO_3$ in the presence of NH_4Cl and NH_4OH.

Group VI—No specific group reagent.

Q. 60. Why is it essential to add dil. HCl before proceeding to the test for the basic radicals of group II?

Ans. Dilute HCl increases the concentration of H^+ ions in the solution and hence suppresses the dissociation of H_2S due to common ion effect. As a result of which the sulphide ion concentration is sufficient only to exceed the solubility products of the sulphides of group II cations.

Since the solubility products (Ksp) for the sulphides of groups III and IV cations are very high, those cations are not precipitated under the above conditions.

Q. 61. Why is the O.S. boiled with conc. HNO_3 in group III?

Ans. In the presence of NH_4Cl, $Fe(OH)_2$ is not completely precipitated because of its high solubility product. For this reason Fe^{2+} salts are oxidised to Fe^{3+} salts by boiling with conc. HNO_3 before adding NH_4Cl and NH_4OH; otherwise Fe^{2+} would not be completely precipitated in III group.

Q. 62. Why is NH_4Cl added along with NH_4OH in group III?

Ans. It is done in order to decrease the concentration of OH^- ions by suppressing the ionisation of NH_4OH by common ion effect. If NH_4OH alone is used in that case, the concentration of OH^- is large enough to ppt. the hydroxide of IV, V and VI cations.

Q. 63. What is blue lake?

Ans. It is blue particles (blue litmus adsorbed on white ppt. of $Al(OH)_3$ floating in colourless solution.

Q. 64. H_2S gas is passed in presence of NH_4OH in group IV. Explain why?

Ans. When H_2S gas is passed in alkaline medium or NH_4OH, the H^+ ions from the dissociation of H_2S gas combine with hydroxyl ions (OH^-) from the dissociation of NH_4OH to form nearly unionised H_2O.

$$H_2S \rightleftharpoons 2H^+ + S^{2-}$$

$$2NH_4OH \rightleftharpoons 2OH^- + 2NH_4^+$$

$$H^+ + OH^- \rightleftharpoons H_2O$$

The removal of H^+ ions from the solution causes more of H_2S to dissociate, thereby increasing the concentration of S^{2-} ions to such an extent that the ionic product of IV group metal sulphides exceeds their solubility product. Hence they are precipitated.

Q. 65. Presence of NH_4Cl is quite essential before the addition of $(NH_4)_2\ CO_3$ in group V. Explain why?

Ans. Ammonium chloride suppresses the ionisation of NH_4OH and $(NH_4)_2CO_3$ due to common ion effect which results in the decrease in the concentration of OH^- and CO_3^{2-} ions. So the ionic product does not exceed the solubility product of $Mg(OH)_2$ or $MgCO_3$ and thereby they are not precipitated during analysis of group V cations.

Q. 66. Na_2CO_3 cannot be used in place of $(NH_4)_2\ CO_3$ in the group V. Explain why?

Ans. Na_2CO_3 is highly ionised electrolyte, which produces very high conc. of CO_3^{2-} ions. As a result ionic product of $MgCO_3$ may exceed its Ksp and it may get precipitated along with the radicals of V group.

Q. 67. An aqueous solution of HCl has conc. 10^{-8} M. What is the approximate value of pH of this solution?

Ans. Slightly less than 7.

Q. 68. How will you prepare chlorine water?

Ans. Take conc. HCl in a test tube and add $KMnO_4$ solution dropwise till the pink colour starts persisting. Now add a few drops of conc. HCl so that pink colour disappears. The colourless solution thus obtained is chlorine water.

Q. 69. Can we use ammonium sulphate in place of ammonium chloride in precipitation of group III cations?

Ans. No, ammonium sulphate cannot be used because it would cause precipitation of group V radicals as their sulphates during analysis of group III.

Q. 70. Name a cation which is not obtained from a metal.

Ans. Ammonium ion (NH_4^+).

INVESTIGATORY PROJECTS

INVESTIGATORY PROJECT 1

Study of Oxalate Ion Content in Guava Fruit

INTRODUCTION

Guava is sweet, juicy and light or dark green coloured fruit. It is cultivated in all parts of India. When ripe it acquires yellow colour and has a penetrating strong scent. The fruit is rich in vitamin C and minerals. It is a rich source of oxalate and its content in the fruit varies during different stages of ripening.

OBJECTIVE OF PROJECT

In this project, we will learn to test for the presence of oxalate ions in the guava fruit and how its amount varies during different stages of ripening.

EXPERIMENT 1

AIM: ***To study the presence of oxalate ion content in guava fruit at different stages of ripening.***

REQUIREMENTS

100 ml measuring flask, pestle and mortar, beaker (250 ml), titration flask, funnel, burette, weight-box, pipette, filter paper, dilute H_2SO_4, 0.05 N $KMnO_4$ solution, guava fruits at different stages of ripening.

THEORY

Oxalate ions are extracted from the fruit by boiling pulp with dil. H_2SO_4. Then oxalate ions are estimated volumetrically by titrating the solution with standard $KMnO_4$ solution.

PROCEDURE

1. Weigh 50.0 g of fresh guava and crush it to a fine pulp using pestle-mortar.
2. Transfer the crushed pulp to a beaker and add about 50 ml dil. H_2SO_4 to it. Boil the contents for about 10 minutes.
3. Cool and filter the contents in a 100 ml measuring flask. Make the volume upto 100 ml by adding distilled water.
4. Take 20 ml of the solution from the measuring flask into a titration flask and add 20 ml of dilute suphuric acid to it. Heat the mixture to about 60°C and titrate it against N/20 $KMnO_4$ solution taken in a burette. The end point is appearance of permanent light-pink colour.
5. Repeat the above experiment with 50.0 g of 1, 2 and 3 days old guava fruit.

OBSERVATIONS

Weight of guava fruit taken each time = 50.0 g

Volume of guava extract taken in each titration = 20.0 ml.

Normality of $KMnO_4$ solution $= \frac{1}{20}$.

Guava extract from	*Burette readings*		*Concordant volume of N/20 $KMnO_4$ solution used*
	Initial	*Final*	
Fresh guava			x ml
One day old guava			x_1 ml
Two days old guava			x_2 ml
Three days old guava			x_3 ml

CALCULATIONS

For fresh guava

$$\underset{\text{(guava extract)}}{N_1V_1} = \underset{(KMnO_4 \text{ soln.})}{N_2V_2}$$

$$N_1 \times 20 = \frac{1}{20} \times x$$

Normality of oxalate, $N_1 = \frac{x}{400}$

Strength of oxalate in fresh guava extract

= Normality × Eq. mass of oxalate ion

$= \frac{x}{400} \times 44$ g/litre of the diluted extract.

Similarly, calculate the strength of oxalate in 1, 2 and 3 days old guava extract and interpret the result.

CONCLUSION

The strength of oxalate ions ...... (increases/decreases) as the guava fruit ripens.

INVESTIGATORY PROJECT 2

Study of the Quantity of Caesin Present in Different Samples of Milk

INTRODUCTION

Milk is a complete diet as it contains in it proteins, carbohydrates, fats, minerals, vitamins and water. Average composition of milk from different sources is given below:

Source of milk	*Water (per cent)*	*Minerals (per cent)*	*Proteins (per cent)*	*Fats (per cent)*	*Carbohydrates (per cent)*
Cow	87.1	0.7	3.4	3.9	4.9
Human	87.4	0.2	1.4	4.0	4.9
Goat	87.0	0.7	3.3	4.2	4.8
Sheep	82.6	0.9	5.5	6.5	4.5

Caesin is the major protein constituent present in the milk and is a mixed phosphoprotein. Caesin has isoelectric pH of about 4.7 and can be easily separated around this pH. It readily dissolves in dilute acids and alkalies. Caesin is present in milk as calcium caseinate in the form of *micelles*. These micelles have negative charge and on adding acid to milk the negative charges are neutralized.

$$Ca^{2+}\text{—Caesinate} + 2CH_3COOH(aq) \longrightarrow \text{Caesin}(s) + (CH_3COO)_2Ca(aq).$$

OBJECTIVE OF PROJECT

The aim of this project is to determine the amount of *caesin* present in different samples of milk.

EXPERIMENT 1

AIM: ***To study the quantity of caesin present in different samples of milk.***

REQUIREMENTS

Beakers (250 ml), filter paper, glass rod, weight box, filtration flask, buchner funnel, water pump, test tubes, porcelain dish, burner, different samples of milk, 1% acetic acid solution, saturated ammonium sulphate solution.

PROCEDURE

1. Take a clean dry beaker, put into it 20 ml of cow's milk and add 20 ml of saturated ammonium sulphate solution slowly and with stirring. Fat along with caesin will precipitate out.

2. Filter the solution and transfer the precipitates in another beaker. Add about 30 ml of water to the precipitates. Only caesin dissolves in water forming milky solution leaving fat undissolved.
3. Heat the milky solution to about 40°C and add 1% acetic acid solution dropwise, when caesin gets precipitated.
4. Filter the precipitate, wash with water, and let them dry.
5. Weigh the dry solid mass in a previously weighed watch glass.
6. Repeat the experiment with other samples of milk.

OBSERVATIONS

Volume of milk taken in each case = 20 ml.

Sample of Milk	*Weight of Caesin*	*% of Caesin*
Cow milk		
Buffalo milk		
Goat milk		
Sheep milk		

CONCLUSIONS

Different samples of milk contain different percentage of caesin.

INVESTIGATORY PROJECT 3

Preparation of Soyabean Milk and its Comparison with Natural Milk

INTRODUCTION

Natural milk is an opaque white fluid secreted by the mammary glands of female mammal. The main constituents of natural milk are proteins, carbohydrates, minerals, vitamins, fats and water and is a complete balanced diet. Fresh milk is sweetish in taste. However, when it is kept for a long time at a temperature of 35 ± 5°C it becomes sour because of bacteria present in air. These bacteria convert lactose of milk into lactic acid which is sour in taste. In acidic conditions caesin of milk starts separating out as a precipitate. When the acidity in milk is sufficient and temperature is around 36°C, it forms semi-solid mass, called curd.

Soyabean milk is made from soybeans. It resembles natural milk. The main constituents of soyabean milk are *proteins, carbohydrates, fats, minerals and vitamins*. It is prepared by keeping soyabeans dipped in water for sometime. The swollen soyabeans are then crushed to a paste which is then mixed with water. The solution is filtered and filtrate is soyabean milk.

EXPERIMENT 1

AIM: ***Preparation of soyabean milk and its comparison with the natural milk with respect to curd formation, effect of temperature and taste.***

REQUIREMENTS

Beakers, pestle and mortar, measuring cylinder, a spoon, tripod stand, thermometer, muslin cloth and burner.

Soyabeans, buffalo milk, fresh curd and distilled water.

PROCEDURE

1. Soak about 150 g of soyabeans in sufficient amount of water so that they are completely dipped in it. Keep them dipped for 24 hours.
2. Take out swollen soyabeans and grind them to a very fine paste with a pestle-mortar.
3. Add about 250 ml of water to this paste and filter it through a muslin cloth. Clear white filtrate is soyabean milk. Compare its taste with buffalo milk.
4. Take 50 ml of buffalo milk in each of the three beakers (labelled as 1, 2, and 3) and heat the beakers to 30°, 40° and 50°C respectively. Add 1/4 spoonfull curd to each of the beakers. Mix well with a spoon and leave the beakers undisturbed for 8 hours and curd is ready.

5. Similarly, take 50 ml of soyabean milk in each of the three other beakers (labelled as 4, 5, 6) and heat the beakers to 30°, 40° and 50°C respectively. Add 1/4 spoonfull curd to each of these beakers. Mix well with a spoon and leave the beakers undisturbed for 8 hours and curd is formed.

OBSERVATIONS

Type of milk	*Beaker No.*	*Temperature*	*Quality of curd*	*Taste of curd*
Buffalo milk	1	30°C		
	2	40°C		
	3	50°C		
Soyabean milk	4	30°C		
	5	40°C		
	6	50°C		

RESULT

For buffalo milk, the best temperature for the formation of good quality and tasty curd is.... °C and for soyabean milk, it is .....°C.

INVESTIGATORY PROJECT 4

Study of Effect of Potassium Bisulphite as Food Preservative Under Various Conditions

INTRODUCTION

Growth of micro-organisms in a food material can be inhibited by adding certain chemical substance. However the chemical substances should not be harmful to the human beings. **Such chemical substances which are added to food materials to prevent their spoilage are known as chemical preservatives.** In our country, two chemical preservatives which are permitted for use are:

1. *Benzoic acid* (or sodium benzoate)
2. *Sulphur dioxide* (or potassium bisulphite)

Benzoic acid or its sodium salt, sodium benzoate is commonly used for the preservation of food materials. For the preservation of fruits, fruit juices, squashes and jams, sodium benzoate is used as preservative because it is soluble in water and hence easily mixes with the food product.

Potassium bisulphite is used for the preservation of colourless food materials such as fruit juices, squashes, apples and raw mango *chutney*. This is not used for preserving coloured food materials because sulphur dioxide produced from this chemical is a bleaching agent. Potassium bisulphite on reaction with acid of the juice liberates sulphur dioxide which is very effective in killing the harmful micro-organisms present in food and thus prevents it from getting spoiled.

$$HSO_3^-(aq) + H^+(aq) \longrightarrow H_2O(l) + SO_2(g)$$

The advantage of this method is that no harmful chemical is left in the food.

The aim of the project is to study the effect of potassium bisulphite as food preservative.

(*i*) at different temperatures.

(*ii*) at different concentrations and

(*iii*) for different intervals of time.

EXPERIMENT 1

AIM: ***To study the effect of potassium bisulphite as food preservative under various conditions (concentration, time and temperature).***

REQUIREMENTS

Conical flasks (100 mL) a mixer, glass rod, knife, apples, sugar and potassium bisulphite

PROCEDURE

1. Take 500 g fresh apples. Wash them thoroughly and peel off the outer layer. Remove the seeds and crush the apples in a mixer. Add about 100 g of sugar and heat the contents slowly for about 10 minutes to prepare jam. During heating keep on stirring the contents. Use this jam for performing the following experiments.

Study of Effect of conc. of Potassium Bisulphite and the Effect of Time

1. Take four conical flasks and label them as A, B, C and D. Add 50 g of jam in each of the four conical flasks.
2. To flask A add 0.1 g, flask B 0.2 g, flask C 0.5 g and flask D 1.0 g of potassium bisulphite. Mix the contents in each flask and leave them undisturbed at room temperature.
3. For some days check for any growth of micro-organisms after each day and record the observations in a table.

OBSERVATIONS

Sample	*Amount of jam*	*Amount of potassium bisulphite*	*Growth of micro-organisms after*					
			1 day	*2 days*	*3 days*	*4 days*	*5 days*	*6 days*
A	50 g	0.1 g						
B	50 g	0.2 g						
C	50 g	0.5 g						
D	50 g	1.0 g						

RESULT

As the concentration of potassium bisulphite is increased, the growth of micro-organisms appears after more days (longer period). The minimum concentration of potassium bisulphite required for preserving jam is approximately 1%.

Once the micro-organisms appear their growth increases with the passage of time.

Study of Effect of Temperature

1. Take three conical flasks and label them as A, B and C. Add 50 g of jam in each of the three flasks.
2. Add 0.5 of potassium bisulphite to each of the three conical flasks. Stir the contents with the help of glass rod to affect thorough mixing.
3. Keep flask A in a refrigerator, flash B at room temperature and flask C in an oven maintained at a temperature of 60°C, leave them undisturbed for few days.
4. Check for any growth of micro-organisms after each day and record the observations.

OBSERVATIONS

Sample	*Amount of jam*	*Amount of Potassium bisulphite*	*Temperature*	*Growth of micro-organisms after*		
				5 days	*10 days*	*15 days*
A	50 g	0.5 g	0°—5° C			
B	50 g	0.5 g	25°—30°C			
C	50 g	0.5 g	60°—70°C			

RESULT

The growth of micro-organisms occurs earliest in the flask kept at room temperature. The preservation of jam by potassium bisulphite is maximum at lower temperature (0°–5°C).

INVESTIGATORY PROJECT 5

Comparative Study of the Rate of Fermentation of Various Food Materials

INTRODUCTION

Fermentation is *the slow decomposition of complex organic compounds into simpler compounds by the action of enzymes.* **Enzymes** *are complex organic compounds, generally proteins.* There are many examples of fermentation processes which we come across in daily life; souring of milk or curd, bread making, wine making and brewing. Fermentation word has been derived from Latin (*Ferver* which means 'to boil'). As during fermentation there is lot of frothing of the liquid due to the evolution of carbon dioxide, it gives the appearance as if it is boiling.

Louis Pasteur in 1860 demonstrated that fermentation is a purely physiological process carried out by living micro-organisms like yeast. This view was abandoned in 1897 when Buchner demonstrated that yeast extract could bring about alcoholic fermentation in the absence of any yeast cells. He proposed that fermenting activity of yeast is due to active catalysts of biochemical origin. These biochemical catalysts are called enzymes. Enzymes are highly specific. A given enzyme acts on a specific compound or a closely related group of compounds.

Sugars like glucose and sucrose when fermented in the presence of yeast cells are converted to ethyl alcohol.

Sucrose is first converted to glucose and fructose with an enzyme *invertase*. Enzyme *zymase* converts glucose and fructose to ethanol.

$$\underset{\text{Sucrose}}{C_{12}H_{22}O_{11}} + H_2O \xrightarrow{\textit{Invertase}} \underset{\text{Glucose}}{C_6H_{12}O_6} + \underset{\text{Fructose}}{C_6H_{12}O_6}$$

$$\underset{\text{Glucose or fructose}}{C_6H_{12}O_6} \xrightarrow{\textit{Zymase}} \underset{\text{Ethyl alcohol}}{2C_2H_5OH} + 2CO_2$$

During fermentation of starch, starch is first hydrolysed to maltose by the action of enzyme *diastase*. The enzyme *diastase* is obtained from germinated barley seeds. Maltose is converted to glucose by enzyme *maltase*. Glucose is converted to ethanol by another enzyme *zymase*.

$$\underset{\text{Starch}}{2(C_6H_{10}O_5)_n} + nH_2O \xrightarrow{\textit{Diastase}} \underset{\text{Maltose}}{nC_{12}H_{22}O_{11}}$$

$$\underset{\text{Maltose}}{C_{12}H_{22}O_{11}} + H_2O \xrightarrow{\textit{Maltase}} \underset{\text{Glucose}}{2C_6H_{12}O_6}$$

$$C_6H_{12}O_6 \xrightarrow{\text{Zymase}} 2C_2H_5OH + 2CO_2$$
Glucose → Ethyl alcohol

Enzymes maltase and zymase are obtained from yeast.

EXPERIMENT 1

AIM: *To compare the rates of fermentation of the following fruit or vegetable juices (i) Apple juice (ii) Orange juice (iii) Carrot juice.*

THEORY

The fruit and vegetable juices contain sugars such as sucrose, glucose and fructose. These sugars on fermentation in the presence of the enzymes *invertase* and *zymase* give ethanol with the evolution of carbon dioxide.

$$\underset{\text{Sucrose}}{C_{12}H_{22}O_{11}} \xrightarrow{\text{Invertase}} \underset{\text{Glucose}}{C_6H_{12}O_6} + \underset{\text{Fructose}}{C_6H_{12}O_6}$$

$$\underset{\text{Glucose}}{C_6H_{12}O_6} + \underset{\text{Fructose}}{C_6H_{12}O_6} \xrightarrow{\text{Zymase}} \underset{\text{Ethanol}}{2C_2H_5OH} + 2CO_2\uparrow$$

Glucose and fructose are reducing sugars and give red coloured precipitates with Fehlings solution, when warmed. When the fermentation is complete, the reaction mixture stops giving any red colour or precipitate with Fehling solution.

REQUIREMENTS

Conical flasks (250 mL), test tubes and water bath, Apple juice, Orange juice, Carrot juice, Fehling solution A, Fehling solution B, solution of Pasteur salts and distilled water.

PROCEDURE

1. Take 5.0 mL of apple juice in a clean 250 mL conical flask and dilute it with 50 mL of distilled water.
2. Add 2.0 g of Baker's yeast and 5.0 mL of solution of Pasteur's salts to the above conical flask.
3. Shake well the contents of the flask and maintain the temperature of the reaction mixture between 35°–40°C.
4. After 10 minutes take 5 drops of the reaction mixture from the flask and add to a test tube containing 2 mL of Fehling reagent.

   Place the test tube in boiling water bath for about 2 minutes and note the colour of the solution or precipitate.
5. Repeat the step 4 after every 10 minutes. When the reaction mixture stops giving any red colour or precipitate with Fehling reagent, the completion of fermentation is indicated.
6. Note the time taken for completion of fermentation.
7. Repeat the above experiment by taking 5.0 mL of carrot juice.

Pasteur's Salt Solution

Pasteur salt solution is prepared by dissolving ammonium tartrate, 10.0 g; potassium phosphate 2.0 g; calcium phosphate 0.2 g, and magnesium sulphate, 0.2 g dissolved in 860 ml of water.

OBSERVATIONS

Volume of fruit juice taken = 5.0 mL

Volume of distilled water added = 50.0 mL

Weight of Baker's yeast added = 2.0 g

Volume of solution of Pasteur's salts = 5.0 mL

Time (in minutes)	*Colour of reaction mixture on reaction with Fehling solution in case of*	
	Apple juice	*Carrot juice*
10		
20		
30		
40		
50		
60		
⋮		
⋮		

RESULT

The rate of fermentation of apple juice is ...... than the rate of fermentation of carrot juice.

EXPERIMENT 2

AIM: ***To compare the rates of fermentation of the given samples of wheat flour, gram flour, rice and potatoes.***

THEORY

Wheat flour, gram flour, rice and potatoes contain starch as the major constituent. Starch present in these food materials is first brought into solution. In the presence of enzyme *diastase* starch undergoes fermentation to give maltose.

$$\underset{\text{Starch}}{2(C_6H_{10}O_5)_n} + nH_2O \xrightarrow{\textit{Diastase}} \underset{\text{Maltose}}{nC_{12}H_{22}O_{11}}$$

Starch gives blue-violet colour with iodine, whereas products of fermentation of starch do not give any characteristic colour. When the fermentation is complete the reaction mixture stops giving blue-violet colour with iodine solution. By comparing the time required for completion of fermentation of equal amounts of different substances containing starch, their rates of fermentation can be compared. The enzyme *diastase* is obtained by germination of moist barley seeds in dark at 15°C. When the germination is complete, the temperature is raised to 60°C to stop further growth. These seeds are crushed in water and filtered. The filtrate contains enzyme diastase and is called malt extract.

REQUIREMENTS

Conical flasks, test tubes, funnel, filter paper and water bath. Wheat flour, gram flour, rice flour, potatoes, 1% iodine solution.

PROCEDURE

1. Take 5.0 g of wheat flour in a 100 mL conical flask and add 30 mL of distilled water.
2. Boil the contents of the flask for about 5 minutes.
3. Filter the above contents after cooling. The filtrate obtained is wheat flour extract.
4. To the wheat flour extract taken in a conical flask, add 5 mL of 1% aqueous NaCl solution.
5. Keep this flask in a water bath maintained at a temperature of 50–60°C. Add 2 mL of malt extract.
6. After 2 minutes take 2 drops of the reaction mixture and add to diluted iodine solution. Note the colour produced.
7. Repeat step 6 after every 2 minutes. When no bluish colour is produced the fermentation is complete. Record the total time taken for completion of fermentation.
8. Repeat the experiment with gram flour extract, rice flour extract, potato extract and record the observations.

OBSERVATIONS

Time (in minutes)	*Colour of the reaction mixture obtained with iodine solution in case of*			
	Wheat flour extract	*Gram flour extract*	*Rice flour extract*	*Potato extract*
2 4 6 8 ⋮ ⋮				

RESULT

The rate of fermentation of starch in different substances containing starch is in the order ....

INVESTIGATORY PROJECT 6

Extraction of Essential Oils Present in *Saunf* (Aniseed), *Ajwain* (Carum) and *Illaichi* (Cardamom)

INTRODUCTION

We are all familiar with the pleasant odours coming out from flowers, spices and many trees. The essences or aromas of plants are due to volatile oils present in them. These smelling volatile oils present in plants are called **essential oils.** Cinnamon, clove, cumin, eucalyptus, garlic, jasmine, peppermint, rose, sandalwood, spearmint, thyme, wintergreen are a few familiar examples of valuable essential oils. The term essential oils literally means "*oils derived from the essence*" of plants.

Essential oil are mainly used for their pleasant odours and flavors in perfumes and as flavoring agents in foods. Some are used in medicines (*e.g.*, camphor, wintergreen, eucalyptus) others as insect repellants (*e.g.*, citronella). Chemically essential oils are composed of complex mixtures of esters, elcohols, phenols, aldehydes, ketones and hydrocarbons. They are essentially non-polar compounds and are thus soluble in non-polar solvents such as petroleum ether, benzene etc. Essential oils may occur in all parts of the plant, but they are often concentrated in the seeds or flowers. They are obtained from the plants by the process of steam distillation and extraction. The technique of steam distillation permits the separation of volatile components from non-volatile materials without raising the temperature of the distillation above 100°C.

Thus, steam distillation reduces the risk of decomposition of essential oils.

EXPERIMENT 1

AIM: ***To extract essential oils present in Saunf (Aniseed), Ajwain (Carum) and Illaichi (Cardamom).***

REQUIREMENTS

Steam generator (Copper Vessel), round bottom flask (500 mL), conical flask, condenser, glass tubes, iron stand, sand bath, separatory funnel, tripod stands, burners, *Ajwain*, *Saunf*, *Illaichi*, petroleum ether (60–80°C)

PROCEDURE

1. Set the apparatus as shown in Fig. P-6.1. The apparatus consists of a steam generator connected to the round bottom flask through a glass inlet tube. The flask is connected to a water condenser through a glass outlet tube. Condenser is further attached to a receiver through a adaptor.
2. Take about 750 ml of water in the steam generator and start heating to produce steam.

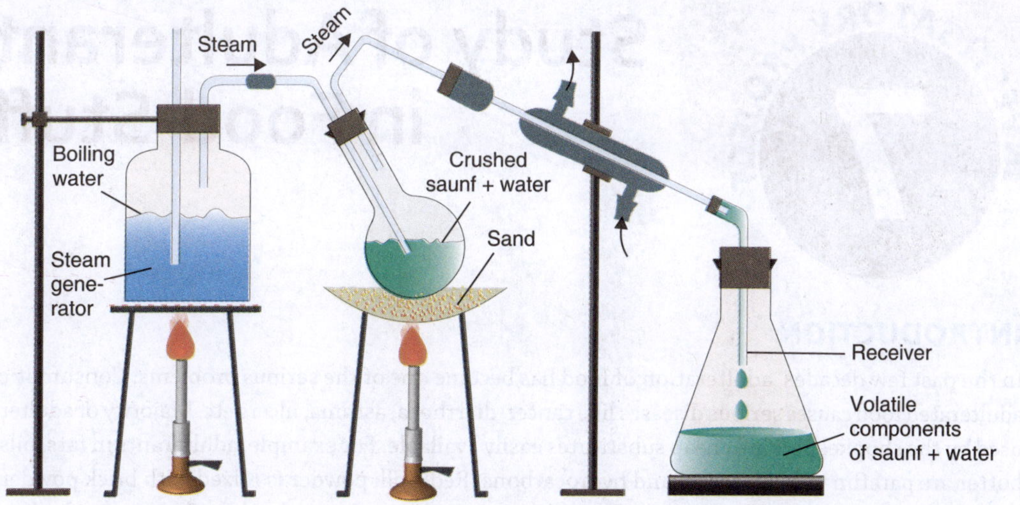

Fig. P-6.1. Extraction of essential oil by steam distillation.

3. In the round bottom flask take 75 g of crushed *saunf*.
4. A vigorous current of steam from steam generator is passed through the round bottom flask.
5. A part of the steam condenses in the round bottom flask. As more and more steam is passed, the steam volatile components of saunf pass through the condenser along with steam. These contents on condensation are collected in the receiver.
6. The contents in the rounds bottom flask may be heated by a bunsen burner to prevent excessive condensation of steam.
7. The process of steam distillation is continued for about half an hour.
8. Transfer the distilate to a separating funnel and extract with 20 mL portions of petroleum ether 3 times.
9. Combine the petroleum ether extracts in a 250 mL conical flask and dry it with the help of anhydrous sodium sulphate.
10. Remove the solvent from the dried filtrate by careful distillation on a water bath. The essential oil is left behind in the distillation flask.
11. Find the weight of the extracted essential oil. Note the colour, odour and weight of the essential oil.
12. Repeat the experiment with *ajwain* and *illaichi*.

OBSERVATIONS

Weight of *saunf* taken	= 75 g
Initial weight of the bottle	= .... g (say x g)
Weight of the bottle + essential oil	= .... g (say y g)
Weight of the essential oil extracted	= .... g ($y - x$ g)
Percentage of essential oil	$= \frac{y-x}{75} \times 100$
Colour of the oil	= ....
Odour of the oil	= ....

Similarly, record the observations of *ajwain* and *illaichi*.

INVESTIGATORY PROJECT 7

Study of Adulterants in Food-Stuffs

INTRODUCTION

In the past few decades, adulteration of food has become one of the serious problems. Consumption of adulterated food causes serious diseases like cancer, diarrhoea, asthma, ulcers etc. Majority of adulterants used by the shopkeepers are cheap substitutes easily available. For example, adulterants in fats, oils and butter are paraffin wax, castor oil and hydrocarbons. Red chilli powder is mixed with brick powder and pepper is mixed with dried papaya seeds. These adulterants can be easily identified by simple chemical tests.

OBJECTIVE OF PROJECT

The aim of this project is to study some of the common food adulterants present in different food-stuffs.

EXPERIMENT 1

AIM: ***To detect the presence of adulterants in fat, oil and butter.***

REQUIREMENTS

Test tube, conc. HCl, furfural, acetic anhydride, conc. H_2SO_4, acetic acid, conc. HNO_3.

PROCEDURE

Common adulterants present in ghee and oil are paraffin wax, hydrocarbons, dyes and argemone oil. These are detected as follows:

(i) Adulteration of vegetable ghee in desi ghee **(Bandouin test)**

Take small amount of desi ghee in a test tube and add to it 1 ml of HCl and 2–3 drops of 2% alcoholic solution of furfural. Shake the contents vigorously. Appearance of red colour in the acid layer shows that vegetable ghee has been mixed as an adulterant to desi ghee.

(ii) Adulteration of paraffin wax and hydrocarbon in vegetable ghee

Heat small amount of vegetable ghee with acetic anhydride. Droplets of oil floating on the surface of unused acetic anhydride indicates the presence of wax or hydrocarbon.

(iii) Adulteration of dyes in fat

Heat 1 ml of fat with a mixture of 1 ml of conc. sulphuric acid and 4 ml of acetic acid. Appearance of pink or red colour indicates presence of dye in fat.

(iv) Adulteration of argemone oil in edible oils

To small amount of oil in a test tube, add few drops of conc. HNO_3 and shake. Appearance of red colour in the acid layer indicates presence of argemone oil.

EXPERIMENT 2

AIM: ***To detect the presence of adulterants in sugar.***

REQUIREMENTS

Test tubes, conc. H_2SO_4, alcoholic solution of α-naphthol and dil HCl.

PROCEDURE

Sugar is usually contaminated with washing soda and other insoluble substances which are detected as follows:

(i) Adulteration of various insoluble substances in sugar

Take small amount of sugar in a test tube and shake it with little water. Pure sugar dissolves in water but insoluble impurities do not dissolve.

(ii) Adulteration of chalk powder, washing soda in sugar

To small amount of sugar in a test tube, add a few drops of dil. HCl. Brisk effervescence of CO_2 shows the presence of chalk powder or washing soda in the given sample of sugar.

EXPERIMENT 3

AIM: ***To detect the presence of adulterants in samples of chilli powder, turmeric powder and pepper.***

REQUIREMENTS

Test tubes, conc. HCl, dil. HNO_3 and KI solution.

PROCEDURE

Common adulterants present in chilli powder, turmeric powder and pepper are red coloured lead salts, yellow lead salts and dried papaya seeds respectively. They are detected as follows:

(i) Adulteration of red lead salts in chilli powder

To a sample of chilli powder add dil. HNO_3. Filter the solution and add 2 drops of potassium iodide solution to the filtrate. Appearance of yellow ppt. indicates the presence of lead salts in chilli powder.

(ii) Adulteration of yellow lead salts to turmeric powder

To a sample of turmeric powder add conc. HCl. Appearance of majenta colour shows the presence of yellow oxides of lead in turmeric powder.

(iii) Adulteration of brick powder in red chilli powder

Add small amount of given red chilli powder in beaker containing water. Brick powder settles at the bottom while pure chilli powder floats over water.

(iv) Adulteration of dried papaya seeds in pepper

Add small amount of sample of pepper to a beaker containing water and stir with a glass rod. Dried papaya seeds being lighter float over water while pure pepper settles at the bottom.

INVESTIGATORY PROJECT 8

Preparation of an Alum from Scrap Aluminium

INTRODUCTION

Aluminium because of its low density, high tensile strength and resistance to corrosion is widely used for the manufacture of airplanes, automobiles lawn furniture as well as for aluminium cans. Being good conductor of electricity it is also used for transmission of electricity. Aluminium foil is used for wrapping cigarettes, confectionery etc. Aluminium is also used for making utensils. The recycling of aluminium cans and other aluminium products is a very positive contribution to saving our natural resources. Most of the recycled aluminium is melted and recast into other aluminium metal products or used in the production of various aluminium compounds, the most common of which are the alums. Alums are double sulphates having general formula $X_2SO_4 . M_2(SO_4)_3 . 24H_2O$

where, X = monovalent cation such as Na^+, K^+, NH_4^+, etc.

M = trivalent cation such as Al^{+3}, Cr^{+3}, Fe^{+3}, etc.

Some important alums and their names are given below:

Potash Alum	$K_2SO_4 . Al_2(SO_4)_3 . 24H_2O$
Soda Alum	$Na_2SO_4 . Al_2(SO_4)_3 . 24H_2O$
Chrome Alum	$K_2SO_4 . Cr_2(SO_4)_3 . 24H_2O$
Ferric Alum	$(NH_4)_2SO_4 . Fe_2(SO_4)_3 . 24H_2O$

Alums are *isomorphous* crystalline solids which are soluble in water.

Potash alum is used in papermaking, in fire extinguishers in food stuffs and in purification of water. Soda alum is used in baking powders and chrome alum is used in tanning leather and water proofing fabrics.

OBJECTIVE OF PROJECT

To prepare a sample of potash alum from scrap aluminium.

EXPERIMENT 1

AIM: ***To prepare potash alum from scrap aluminium.***

REQUIREMENTS

250 ml conical flask, funnel, beaker, scrap aluminium piece, KOH, 6 M H_2SO_4.

THEORY

Aluminium metal is treated with hot aqueous KOH solution. Aluminium dissolves as potassium aluminate, $KAl(OH)_4$, salt.

$$2Al(s) + 2KOH(aq) + 6H_2O(l) \longrightarrow 2KAl(OH)_4(aq) + 3H_2(g)$$

Potassium aluminate solution on treatment with dil. sulphuric acid first gives precipitate of $Al(OH)_3$, which dissolves on addition of small excess of H_2SO_4 and heating.

$$2KAl(OH)_4(aq) + H_2SO_4(aq) \longrightarrow 2Al(OH)_3(s) + K_2SO_4(aq) + 2H_2O(l)$$

$$2Al(OH)_3(s) + 3H_2SO_4(aq) \xrightarrow{\Delta} Al_2(SO_4)_3(aq) + 6H_2O(l)$$

The resulting solution is concentrated to near saturation and cooled. On cooling crystals of potash alum crystallize out.

$$K_2SO_4(aq) + Al_2(SO_4)_3(aq) + 24H_2O(l) \longrightarrow K_2SO_4 . Al_2(SO_4)_3 . 24H_2O(s)$$

PROCEDURE

1. Clean a small piece of scrap aluminium with steel wool and cut it into very small pieces. Aluminium foil may be taken instead of scrap aluminium.
2. Put the small pieces of scrap aluminium or aluminium foil (about 1.00 g) into a conical flask and add about 50 ml of 4 M KOH solution to dissolve the aluminium. The flask may be heated gently in order to facilitate dissolution. Since during this step hydrogen gas is evolved, this step must be done in a well-ventilated area. Continue heating until all of the aluminium reacts. Filter the solution to remove any insoluble impurities and reduce the volume to about 25 ml by heating.
3. Allow the filtrate to cool. Now add slowly 6 M H_2SO_4 until insoluble $Al(OH)_3$ just forms in the solution.
4. Gently heat the mixture until the $Al(OH)_3$ precipitate dissolves. Cool the resulting solution in an ice-bath for about 30 minutes whereby alum crystals separate out. For better results the solution may be left overnight for the crystallisation to continue. In case the crystals do not form the solution may be further concentrated and cooled again.
5. Filter the crystals from the solution using a vacuum pump, wash the crystals with 50/50 ethanol-water mixture. Continue applying the vacuum until the crystals appear dry.
6. Determine the mass of the alum crystals.

OBSERVATIONS

Mass of aluminium metal	= .... g
Mass of potash alum	= .... g
Theoretical yield of potash alum	= .... g
Percent yield	= .... %

INVESTIGATORY PROJECT 9

Study of the Effect of Metal Coupling on the Rusting of Iron

INTRODUCTION

Metals and alloys undergo rusting and corrosion. The process by which some metals when exposed to atmospheric conditions *i.e.*, moist air, carbon dioxide form undesirable compounds on the surface is known as **corrosion.** The compounds formed are usually oxides. Rusting is also a type of corrosion but the term is restricted to iron or products made from it. Iron is easily prone to rusting making its surface rough. Chemically, rust is a hydrated ferric oxide $Fe_2O_3.nH_2O$. Rusting may be explained by an electrochemical mechanism. In the presence of moist air containing dissolved oxygen or carbon dioxide, the commercial iron behaves as if composed of small electrical cells. At anode of cell, iron passes into solution as ferrous ions.

$$Fe \longrightarrow Fe^{2+} + 2e^-$$

The electrons from the above reaction move towards the cathode and form hydroxyl ions

$$H_2O + (O) + 2e^- \longrightarrow 2OH^-$$

Under the influence of dissolved oxygen the ferrous ions and hydroxyl ions interact to form rust, *i.e.*, hydrated ferric oxide.

$$2Fe^{2+} + H_2O + (O) \longrightarrow 2Fe^{3+} + 2OH^-$$

$$2Fe^{3+} + 6OH^- \longrightarrow \underset{\text{(Rust)}}{Fe_2O_3.3H_2O} \text{ or } 2Fe(OH)_3$$

If supply of oxygen is limited the corrosion product may be black anhydrous magnetite, Fe_3O_4.

METHODS OF PREVENTION OF CORROSION AND RUSTING

Some of the methods used to prevent corrosion and rusting are:

1. **Barrier Protection.** In this method, a barrier film is introduced between iron surface and atmospheric air. The film is obtained by painting, varnishing, etc.
2. **Galvanization.** The metallic iron is covered by a layer of more active metal such as zinc. The active metal loses electrons in preference to iron.

$$Zn \longrightarrow Zn^{2+} + 2e^-$$

Thus, protecting iron from rusting and corrosion.

OBJECTIVE OF PROJECT

In this project the aim is to investigate effect of metal coupling on the rusting of iron. Metal coupling affects the rusting of iron. If the iron nail is coupled with a more electro-positive metal like zinc, magnesium or aluminium rusting is prevented but if on the other hand, it is coupled with less electro-positive metal like copper, the rusting is facilitated.

EXPERIMENT 1

AIM: ***To study the effect of metal coupling on rusting of iron.***

REQUIREMENTS

Two petridishes, four test tubes, four iron nails, beaker, sand paper, wire gauge. Gelatin, copper, zinc and magnesium strips, potassium ferricyanide solution, phenolphthalein.

PROCEDURE

1. Clean the surface of iron nails with the help of sand paper. Wash them with carbon tetrachloride and dry on filter paper.
2. Wind a clean zinc strip around one nail, a clean copper wire around the second and clean magnesium strip around the third nail. Put all these three and a fourth nail in petridishes so that they are not in contact with each other.
3. **Preparation of agar agar solution.** Heat about 3 g of agar agar in 100 ml of water taken in a beaker until solution becomes clear. Add about 1 ml of 0.1 M potassium ferricyanide solution, 1 ml of phenolphthalein solution and stir well the contents.

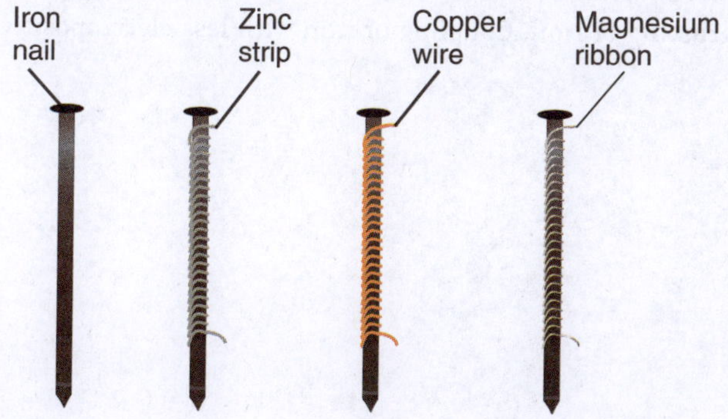

Fig. P-9.1

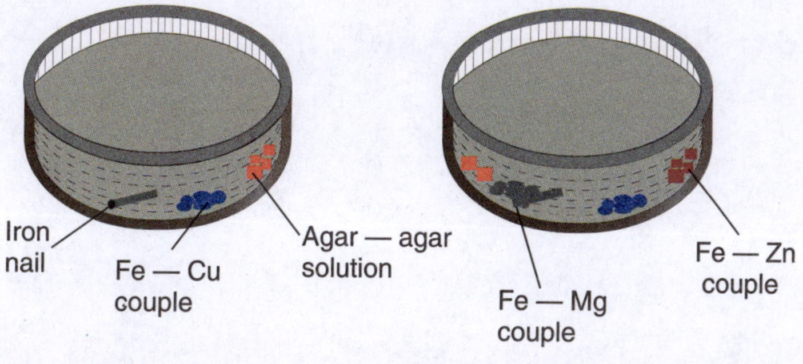

Fig. P-9.2. Study of effect of metal coupling on rusting of iron.

4. Fill the petridishes with hot agar agar solution in such a way that only lower half of the nails are covered with the liquids.
5. Keep the covered petridishes undisturbed for one day or so.
6. The liquid sets to a gel on cooling. Two types of patches are observed around the rusted nail, one is blue and the other pink. Blue patch is due to the reaction between Ferrous ions and potassium ferricyanide to form potassium ferroferricyanide, $KFe[Fe(CN)_6]$ whereas pink patch is due to the formation of hydroxyl ions which turns colourless phenolphthalein to pink.

OBSERVATIONS

S. No.	Metal pair	Colour of the patch	Nail rusts or not
1.	Iron-Zinc		
2.	Iron-Magnesium		
3.	Iron-Copper		
4.	Iron-Nail		

CONCLUSION

It may be concluded that coupling of iron with more electropositive metal such as zinc and magnesium resists corrosion and rusting of iron. Coupling of iron with less electropositive metal such as copper increases rusting.

INVESTIGATORY PROJECT 10

Preparation of Rayon Thread from Filter Paper

INTRODUCTION

Natural fibres are the chief raw materials for the preparation of rayon. The term rayon includes all synthetic fibres obtained from cellulose and are used commercially in fibre manufacturing. Cellulose can be converted into cupra silk, acetate rayon and viscose rayon depending upon the mode of treatment it undergoes while preparing rayon. Among these, viscose rayon is the most common.

Cellulose is an insoluble material which is first converted into a soluble derivative called *viscose*. The viscose is then forced through fine orifices into some reagent (usually dilute sulphuric acid) and the resulting thread is kept under tension to form the fibres of required tensile strength. Rayon, also called artificial silk, is used for manufacturing of fabrics like stockings, shirts, sarees, etc.

OBJECTIVE OF PROJECT

This project is aimed at preparation of rayon thread from filter paper employing cuprammonium process.

EXPERIMENT 1

AIM: ***To prepare rayon threads from filter papers using cuprammonium process.***

REQUIREMENTS

Beakers, conical flasks, filtration flasks, vacuum pump, bent tube, glass rod, 50% ammonia solution, dil. NaOH solution, dil. H_2SO_4, filter paper or waste paper.

PROCEDURE

The cellulose is dissolved in cuprammonium hydroxide $[Cu(NH_3)_4](OH)_2$ and the procedure to be followed is given below:

1. **To Prepare Cuprammonium Hydroxide Solution.** Weigh about 20.0 g of crystalline copper sulphate in a clean watch glass. Dissolve it in 100 ml of water taken in a beaker. Add dilute NaOH solution to this solution slowly with stirring and note the separation of precipitate of $Cu(OH)_2$. Filter the precipitate on water pump and wash the precipitate thoroughly with water so that a portion of filtrate does not indicate the presence of sulphate ions on testing with $BaCl_2$ solution. Now transfer the precipitate to a 250 ml beaker add 50 ml of liquor ammonia. The precipitate will dissolve resulting in a deep blue solution of cuprammonium hydroxide (*Schweitzer's solution*). This is the solvent for dissolving cellulose.
2. **Dissolving the Cellulose Matter.** Weigh about 1 g of ordinary filter paper and cut it into small pieces. Add these pieces to the cuprammonium solution taken in a conical flask. Close the flask

with rubber stopper and allow it to stand for 3-4 days. During this period, filter paper completely dissolves leaving a viscous solution called **viscose.**

3. **Formation of Rayon Filament.** The *viscose* solution is taken in a syringe. Then the nozzle of the syringe is dipped in a 5 M H_2SO_4 solution taken in a wide mouthed beaker. Fig. P-10.1. Squeeze out the viscose into the acid solution and at the same time keep on moving the nozzle in the acid. Long filaments of rayon will be formed in the beaker. The acid bath is left undisturbed for 24 hours, until the blue colour of rayon filament changes to white.

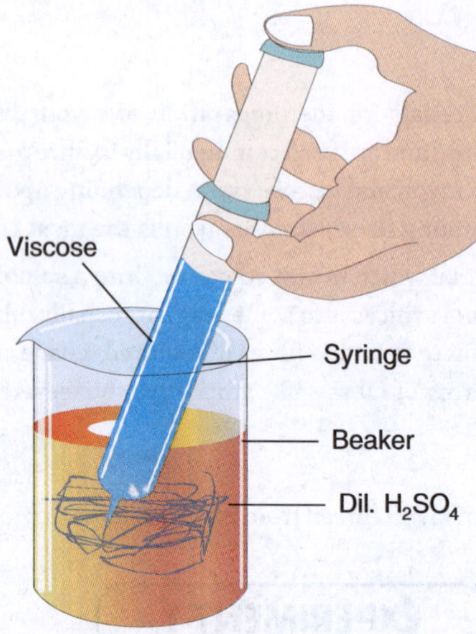

Fig. P-10.1. Formation of rayon filament.

Rayon filaments are then removed from the acid bath, washed with water and dried by keeping them on filter paper. When the threads are completely dried, weigh them and determine the maximum length of the fibre formed.

OBSERVATIONS

Weight of filter paper taken	= ...... g
Weight of rayon filament obtained	= ...... g
Maximum length of the filament	= ...... cm.

INVESTIGATORY PROJECT 11

Dyeing of Fabrics

INTRODUCTION

Dyes *are coloured substances which can adhere to the surface of materials and are used to give colour to paper, food-stuffs and various textiles such as cotton, wool, synthetic fibres, silk etc*. For example, alizarin, indigo, congo red, etc. Chemically, a dye contains:

(*i*) Some group (such as *azo, indigoid, triphenylmethyl, anthraquinone*, etc.) which is responsible for the colour of the dye.

(*ii*) Some group (such as $—NH_2$, $—SO_3H$, $—COOH$, etc.) which makes the dye stick to the fabric by formation of some salt.

The dyed fabrics appear to be coloured because a particular dye absorbs radiations of some specific wavelengths from the visible region of electromagnetic radiations which fall on its surface. The remaining radiations (complementary colours) of light are reflected. The colour which we observe is due to this reflected light. For example, if a dye absorbs the light in the wavelength region corresponding to red, then it would appear green, which is the complementary colour of red. Similarly, if a dye absorbs blue colour, it would appear orange.

CHARACTERISTICS OF A DYE

(*i*) *It must have a suitable colour.*

(*ii*) *It must be capable of being fixed to the material*

(*iii*) *When fixed it must be fast to detergents, soaps, water, dry-cleaning solvents, light and dilute acids.*

TYPES OF DYES

The dyes are classified by dye manufacturers for marketing into the following types:

1. **Acid dyes.** *These are azo dyes and are characterised by the presence of acidic groups*. The presence of acidic group makes the dyes more soluble and serves as the reactive points for fixing the dye to the fibre. They are chiefly used for dyeing wool, silk and nylon. For example, *Orange I* and *Orange II*.
2. **Basic dyes.** *These dyes contain organic basic groups such as NH_2 or NR_2*. In acidic solutions, these form water soluble cations and use the anionic site on the fabric to get themselves attached. These are used for dyeing wool, silk and nylon. For example, *aniline yellow, butter yellow*.
3. **Direct dyes.** These are also azo dyes and are used to dye the fabrics directly by placing in aqueous solution of the dye. These dyes attach to the fabrics by means of hydrogen bonding.

4. **Disperse dyes.** These dyes are applied in the form of a dispersion of minute particles of the dye in a soap solution in the presence of phenol or benzoic acid. These dyes are used to dye rayons, dacron, nylon, polyesters etc. For example, *celliton fast pink B* and *celliton fast blue B*.
5. **Fibre reactive dyes.** These dyes are linked to the fibre by —OH or by —NH_2 group present on the fibre. These dyes induce fast colour on fibres which is retained for a longer time. These dyes are used for dyeing cotton, wool and silk.
6. **Insoluble dyes.** These dyes are directly synthesised on the fibre. The fabric to be coloured is soaked in an alkaline solution of phenol and then treated with a solution of diazotised amine to produce azo dye. The colour induced by such dyes is not fast. These dyes are used for dyeing of cotton, silk, polyester nylon, etc. For example, *nitroaniline red*.
7. **Vat dyes.** These dyes are water-insoluble and before dyeing these are reduced to colourless compounds in wooden vats by alkaline reducing agent. The fibre is then soaked in the solution of the dye. Fibre is then exposed to air or an oxidizing agent. By doing so the colourless compound gets reoxidised to coloured dye on the fabric. For example, *indigo*.
8. **Mordant dyes.** These dyes are applied after treating the fabric with precipitates of certain substances (mordant material) which then combines with the dye to form a coloured complex called lake. Some of the mordants are salts of aluminium, iron and tannic acids. Depending on the mordant used, the same mordant dye can give different colour and shades. For example, *alizarin* gives red colour with aluminium and black violet with iron mordant. Mordant dyes are used for dyeing of wool, silk and cotton.

EXPERIMENT 1

AIM: ***To dye wool and cotton clothes with malachite green.***

REQUIREMENTS

500 ml beakers, tripod stand, wire gauze, glass rod, spatula, wool cloth and cotton cloth. Sodium carbonate, tannic acid, tartaremetic and malachite green dye.

PROCEDURE

1. **Preparation of sodium carbonate solution.** Take about 0.5 g of solid sodium carbonate and dissolve it in 250 ml of water.
2. **Preparation of tartaremetic solution.** Take about 0.2 g of tartaremetic and dissolve it in 100 ml of water by stirring with the help of glass rod.
3. **Preparation of tannic acid solution.** Take 100 ml of water in a beaker and add about 1.0 g of tannic acid to it. Heat the solution. On heating a clear solution of tannic acid is obtained.
4. **Preparation of dye solution.** Take about 0.1 g of malachite green dye and add to it 400 ml of water. On warming a clear solution of the dye results.
5. **Dyeing of wool.** Take about 200 ml of dye solution and dip in it the woollen cloth to be dyed. Boil the solution for about 2 minutes. After that remove the cloth and wash it with hot water 3-4 times, squeeze and keep it for drying.

6. **Dyeing of cotton.** Cotton does not absorb malachite green readily, therefore it requires the use of a mordant. For dyeing a cotton cloth dip it in sodium carbonate solution for about 10 minutes and then rinse with water. Then put the cloth in hot tannic acid solution for about 5 minutes. Now take out the cloth from tannic acid solution and keep it in tartaremetic solution for about 5 minutes. Remove the cloth and squeeze it with spatula to remove most of the solution. Now place the cloth in boiling solution of the dye for about 2 minutes. Remove and wash the dyed cloth thoroughly with water, squeeze and keep it for drying.
7. **Dyeing of cotton directly.** Take another piece of cotton cloth and put it directly into boiling solution of the dye. Keep it dipped for about 2 minutes. Remove the cloth, wash with water, squeeze and keep it for drying.

   Compare the colour of this cloth with that of dyed by using mordant.

OBSERVATIONS

1. The colour of wool cloth dyed directly by dipping in hot solution of malachite green dye is fast.
2. The colour of cotton cloth dyed directly (without using mordant) by dipping in hot solution of malachite green is not fast to washing and is of low intensity.
3. The colour of cotton cloth dyed indirectly by using mordant and then by dipping in hot solution of malachite green is fast to washing and is of high intensity.

INVESTIGATORY PROJECT 12

Sterilization of Water with Bleaching Powder

INTRODUCTION

Water is the major constituent of all living beings. Water is necessary to sustain all types of life. The water used for drinking purpose by human beings should fulfil the following conditions:

(i) *It should be colourless.*

(ii) *It should not possess any smell.*

(iii) *It should not contain any harmful dissolved salts such as nitrates, nitrites, mercury salts, lead salts.*

(iv) *It should not contain any living organisms such as algae, fungi, bacteria, etc.*

In order to obtain water for drinking purposes, water is first treated with alum whereby clay and other colloidal particles get precipitated. The suspended impurities are then removed by filtration and the clear water obtained is subjected to some suitable treatment to destroy harmful germs and bacteria. These becteria cause many dangerous diseases such as cholera, typhoid, dysentery, etc. *The process of killing the harmful bacteria by some suitable treatment of water* is called **sterilization** or **disinfection of water.** The common sterilizing agents are *chlorine, ozone, bleaching powder, potassium permanganate, chloramine*, etc. Sterilization of water can also be done by simply boiling the water for about 15 minutes. However, this method can be applied only on small scale. In the present context, we shall focus on disinfection of water using bleaching powder.

The chemical action of bleaching powder on germs and bacteria is due to the chlorine which becomes available, when it is added to water.

$$CaOCl_2 + H_2O \longrightarrow Ca(OH)_2 + Cl_2.$$

OBJECTIVE OF PROJECT

The objective of this project is to determine the amount of bleaching powder required for the sterilization of given samples of water.

EXPERIMENT 1

AIM: ***Determination of the dosage of bleaching powder required for sterilization or disinfection of different samples of water.***

REQUIREMENTS

Burette, titration flask, 100 ml graduated cylinder, 250 ml measuring flask, weight box, glazed tile, glass wool.

Bleaching powder, 0.1 N $Na_2S_2O_3$ solution, 10% KI solution, different samples of water, starch solution.

THEORY

1. A known mass of the given sample of bleaching powder is dissolved in water to prepare a solution of known concentration. This solution contains dissolved chlorine, liberated by the action of bleaching powder with water.

$$CaOCl_2 + H_2O \longrightarrow Ca(OH)_2 + Cl_2.$$

2. The amount of chlorine present in the above solution is determined by treating a known volume of the above solution with excess of 10% potassium iodide solution, when equivalent amount of iodine is liberated. The iodine, thus liberated is then estimated by titrating it against a standard solution of sodium thiosulphate, using starch solution as indicator.

$$Cl_2 + 2KI \longrightarrow 2KCl + I_2$$

$$I_2 + 2Na_2S_2O_3 \longrightarrow Na_2S_4O_6 + 2NaI$$

3. A known volume of one of the given samples of water is treated with a known volume of bleaching powder solution. The amount of residual chlorine is determined by adding excess potassium iodide solution and then titrating against standard sodium thiosulphate solution.
4. From the readings in 2 and 3, the amount of chlorine and hence bleaching powder required for the disinfection of a given volume of the given sample of water can be calculated.

PROCEDURE

1. **Preparation of bleaching powder solution.** Weigh accurately 2.5 g of the given sample of bleaching powder and transfer it to a 250 ml conical flask. Add about 100–150 ml of distilled water. Stopper the flask and shake it vigorously. The suspension thus obtained is filtered through glass wool and the filtrate is diluted with water (in a measuring flask) to make the volume 250 ml. The solution obtained is 1% **bleaching powder solution.**
2. Take 10 ml of bleaching powder solution in a stoppered conical flask and add to it 20 ml of 10% KI solution. Stopper the flask and shake it vigorously. Titrate this solution against 0.1 N $Na_2S_2O_3$ solution taken in the burette. When the solution in the conical flask becomes light yellow in colour, add about 2 ml of starch solution. The solution now becomes blue in colour. Continue titrating till the blue colour just disappears. Repeat the titration to get a set of three concordant readings.
3. Take 100 ml of the water sample in a 250 ml stoppered conical flask and add to it 10 ml of bleaching powder solution. Then add 20 ml of KI solution and stopper the flask. Shake vigorously and titrate against 0.1 N $Na_2S_2O_3$ solution using starch solution as indicator as described in step 2.
4. Repeat the step 3 with other samples of water and record the observations.

OBSERVATIONS

Weight of the bleaching powder dissolved to prepare 250 ml of solution = 2.5 g.

Titration I. *Bleaching powder solution against 0.1 N $Na_2S_2O_3$ solution*

Volume of bleaching powder solution taken for each titration = 10.0 ml.

Volume of KI solution added = 20.0 ml.

S. No.	Burette Readings		Volume of 0.1 N $Na_2S_2O_3$ solution used
	Initial	Final	
1.	...	...	... ml
2.	...	...	... ml
3.	...	...	... ml
4.	...	...	... ml

Concordant volume = ... ml (say V_1 ml)

Titration II

Volume of water sample I taken for each titration = 100 ml

Volume of bleaching powder solution added = 10.0 ml

Volume of KI solution added = 20.0 ml.

S. No.	Burette Readings		Volume of 0.1 N $Na_2S_2O_3$ solution used
	Initial	Final	
1.	...	...	... ml
2.	...	...	... ml
3.	...	...	... ml
4.	...	...	... ml

Concordant volume = ... ml (say V_2 ml)

Similarly, record the observations for other samples of water.

CALCULATIONS

Amount of bleaching powder used to disinfect 100 ml of water sample I

$$\equiv (V_1 - V_2) \text{ ml of } 0.1 \text{ N } Na_2S_2O_3 \text{ solution}$$

Now 1 ml of bleaching powder solution contains bleaching powder

$$= \frac{2.5}{250} = 0.01 \text{ g}$$

10 ml of bleaching powder solution $\equiv V_1$ ml of 0.1 $Na_2S_2O_3$

$$\therefore \ 1 \text{ ml of } Na_2S_2O_3 \text{ solution} \equiv \frac{10}{V_1} \text{ ml of bleaching powder solution}$$

Volume of bleaching powder solution required to disinfect 100 ml of water sample I

$$= (V_1 - V_2)\frac{10}{V_1} \text{ ml}$$

$$(V_1 - V_2)\frac{10}{V_1} \text{ ml of bleaching powder solution} = \frac{(V_1 - V_2)\,10}{V_1} \times 0.01 \text{ g of bleaching powder}$$

Amount of bleaching powder required to disinfect 1 litre of water sample I

$$= (V_1 - V_2)\frac{10}{V_1} \times \frac{0.01 \times 1000}{100} \text{ g}$$

$$= \frac{(V_1 - V_2)}{V_1} \text{ g}$$

Similarly, calculate the amount of bleaching powder required to disinfect one litre of other samples of water.

RESULT

Amount of the given sample of bleaching powder required to disinfect one litre of water

Sample I = ... g

Sample II = ... g

Sample III = ... g

INVESTIGATORY PROJECT 13

Setting of Cement

INTRODUCTION

Cement is essentially a finely ground mixture of calcium silicates ($3CaO. SiO_2$) and aluminates ($3CaO. Al_2O_3$) which sets to a hard mass when treated with water. This property makes cement capable of joining rigid masses like bricks, stones, tiles etc. into coherent structures. The cements have property of setting and hardening under water due to certain physico-chemical process and are, therefore, called hydraulic cements. During setting of cement, the physical changes taking place are gel formation and crystallisation and chemical changes are hydration and hydrolysis.

The process of solidification of cement paste involves: (*i*) *setting*, and (*ii*) *hardening*. Setting is stiffening of the original plastic mass into initial gel formation. After setting, hardening starts due to gradual start of crystallisation in the interior of the mass. The strength developed by cement at any time depends upon the amount of gel formed and the extent of crystallisation. A mixture of cement, sand, small pieces of stone (gravel) and water is known as **concrete** and sets to an extremely hard structure.

When cement is used for construction purposes, it is always mixed with sand and little water to make a pasty material called **mortar.** Here cement or lime forms the binding material and function of sand is to prevent shrinking and cracking and to increase the bulk, thereby reducing the cost of the mortar. When cement is used as the binding material it is called **cement mortar** and when lime is used as the binder it is called **lime-mortar**. Sand in addition to its other functions also increases the adhesive qualities of the binding material.

Effect of quality of sand on setting of cement mortar. Sand obtained from different sources has different qualities. For example, *sea sand* obtained from sea contains some unwanted salts and retards the setting of cement and is unsuitable for making mortar. On the other hand, *pit sand* obtained from pits in the soil and *river sand* obtained from river bed are considered excellent for preparing mortar and concrete.

Effect of time on setting of cement mortar. Time has an important role on the strength developed by cement mortar. When a cement sand paste in the ratio 1: 3 in water is allowed to dry, the strength of the solid mass keeps on increasing with increase in the time given for setting. It acquires nearly full strength in 28 days.

OBJECTIVE OF PROJECT

In this project, we will study the setting of mixtures of cement with lime and sand and the effect of relative proportion of their masses, the effect of time and the effect of quality of sand on the strength of cement mortar.

EXPERIMENT 1

AIM: ***To study the setting of mixtures of cement with lime, sand of different qualities, rice husk, fly-ash, etc. (with respect to volume and strength).***

REQUIREMENTS

Beakers, glass rod, weights, small wooden boxes or empty match boxes. Lime, pit sand, river sand, cement, fly-ash, rice husk.

PROCEDURE

1. Prepare the sets of mixtures of various compositions as given in Table 1.
2. Take each of the mixtures in different beakers and prepare their pastes by adding minimum quantity of water.
3. Take 10 empty match-box inner cases and mark them from 1 to 10.
4. Transfer the prepared pastes immediately into the match-boxes and compact them by pressing with hand.
5. Spray water from time to time over the pastes so that they are always moist.
6. Take out the slabs after three days and test for its strength. For that hold a weight of 10 g in your hand at a fixed height (say 50 cm above the ground) and drop the weight on the slab. See if the slab breaks or not. If it does not break then take 20 g weight and drop it from the same height. This way keep on increasing the weight and note down the minimum weight required to break the slab.

OBSERVATIONS

Setting time allowed = 3 days.

Table 1. Effect of various types of sand on strength of slab

S. No.	*Composition of mortar (Ratio by volume of various components)*						*Minimum weight required to break the slab*
	Cement	*River sand*	*Pit sand*	*Lime*	*Fly-ash*	*Rice-husk*	
1.	1	3	—	—	—	—	...... g
2.	1	6	—	—	—	—	...... g
3.	1	—	3	—	—	—	...... g
4.	1	—	6	—	—	—	...... g
5.	1	6	—	—	1	—	...... g
6.	2	9	—	—	1	—	...... g
7.	1	3	—	1	—	—	...... g
8.	1	3	—	2	—	—	...... g
9.	1	1	—	—	—	1	...... g
10.	1	3	—	—	—	2	...... g

CONCLUSION

The relative strengths of various slabs of different mixtures of cement and other component is in the order ....

EXPERIMENT 2

AIM: ***To study the setting of mixtures of cement with sand, lime and fly-ash with respect to time and strength.***

REQUIREMENTS

Same as in Experiment 1.

PROCEDURE

1. Prepare mixtures of the various compositions as given in Table 2.
2. Take each of the mixtures in different beakers and prepare their pastes by adding minimum quantity of water.
3. Take 9 empty match-box inner cases and mark them from 1 to 9.
4. Fill three cases with paste of each composition.
5. Spray water from time to time over the pastes so that they remain moist all the time.
6. After three days take out one slab of each composition and test for their strength by the method described in Experiment 1.
7. Similarly, take out a set of three slabs after 7 days and then after 30 days and test for their strengths.

Table 2. Effect of setting time cement on-strength

S. No.	*Composition of mixture*	*Minimum weight required to break the slab after*		
		3 days	*7 days*	*30 days*
1.	Cement: River sand 1 : 3	...... g	...... g	...... g
2.	Cement: River sand: Fly-ash 2 : 9 : 1	...... g	...... g	...... g
3.	Cement: River sand: Lime 1 : 3 : 1	...... g	...... g	...... g

CONCLUSIONS

The strength of the slab increases with increase in the setting time allowed.

INVESTIGATORY PROJECT 14

Study of Presence of Insecticides and Pesticides in Fruits and Vegetables

INTRODUCTION

In the past decade, there has been a tremendous increase in the yields of various crops to meet the demand of over-growing world population. This great feat has been achieved by adopting new methods of farming and by extensive use of fertilizers and insecticides. **Insecticides** and **pesticides** *are chemicals which when sprayed over the crop protect it from pests*. For example, DDT, BHC, zinc phosphide, mercuric chloride, dinitrophenol, etc. They either kill the insects or prevent their growth. All pesticides are poisonous chemicals and must be used in small quantities and with great care. Pesticides have proved to be effective against variety of insects, weeds and fungi and are respectively called *insecticides, herbidicides* and *fungicides*. Most of the pesticides are non-biodegradable and remain penetrated as such into plants, fruits, and vegetables. From plants, they transfer into animals, birds and human beings who eat these polluted fruits and vegetables. Inside the body they remain accumulated and cause serious health problems. These days preference is given to the use of biodegradable insecticides like vapum, malathion. But even these are also not totally harmfree. Samples of raw food commodities including wheat, pulses, fish, meat, butter, fruits, vegetables show a good amount of insecticide residues. Such findings have aroused the concern of scientists, agricultural administrators and health officials all over the world to put a check over the use of insecticides and to search for non-insecticidal means of pest control.

OBJECTIVE OF PROJECT

In this project, our aim is to study the presence of insecticide/pesticide residues in various fruits and vegetables. Since most of the insecticides/pesticides contain nitrogen, we will test for the presence of these insecticide residues by testing for the presence of nitrogen.

EXPERIMENT 1

AIM: ***To study the presence of insecticides/pesticides (nitrogen-containing) in various fruits and vegetables.***

REQUIREMENTS

Mortar and pestle, beakers, funnel, glass-rod, filter-paper, china-dish, water-bath, tripod stand, fusion-tubes, knife, test-tube.

Samples of various fruits and vegetables, alcohol, sodium metal, ferric chloride solution, ferrous sulphate crystals, distilled water and dil. sulphuric acid.

PROCEDURE

1. Take different kinds of fruits and vegetables and cut them into small pieces separately.
2. Transfer the cut pieces of various fruits and vegetables into the mortar separately and crush them.
3. Take different beakers for each kind of fruits and vegetables and place the crushed fruits and vegetables in these beakers and add 10 ml of alcohol to each of these. Stir well and filter. Collect the filtrate in separate china-dishes.
4. Evaporate the alcohol by heating china-dishes one-by-one over a water bath and let the residue dry in an oven.
5. Heat a small piece of dry sodium in a fusion tube, till it melts. Then add one of the above residues from china-dish to this fusion tube and heat till red-hot. Drop the hot fusion tube in a china-dish containing about 10 ml of distilled water. Break the tube and boil the contents of the china-dish for about 5 minutes. Cool and filter the solution. Collect the filtrate.
6. To the filtrate add 1 ml freshly prepared ferrous sulphate solution and warm the contents. Then add 2-3 drops of ferric chloride solution and acidify with dil.HCl. If a blue or green ppt. or colouration is obtained, it indicates the presence of nitrogen containing insecticide.
7. Repeat the test of nitrogen for residues obtained from other fruits and vegetables and record the observations.

OBSERVATIONS

S. No.	Name of the fruit or vegetable	Test for the presence of nitrogen (positive/negative)	Presence of insecticide/pesticide residue
1.			
2.			
3.			
⋮			
⋮			
⋮			

INVESTIGATORY PROJECT 15

Comparative Study of Commercial Antacids

INTRODUCTION

It is well-known that the food we take undergoes a series of complex reactions within the body which constitute what we call digestion and metabolism. These reactions are catalysed by enzymes which are very specific in their action and can function properly only when the pH of the medium is within a specific range.

Some enzymes require mildly alkaline conditions while others operate only in weakly acidic conditions. Amongst the latter category of enzymes are the enzymes which control the digestion of proteins present in the food as it reaches the stomach. In the stomach dilute hydrochloric acid is secreted and it provides mildly acidic conditions required for the functioning of protein digesting enzymes in the stomach. However, sometimes the stomach begins to secrete an excess of hydrochloric acid. This condition is known as *gastric hyperacidity* and is caused by over eating and highly spiced food. This interferes in the normal process of digestion and leads to acute discomfort due to indigestion.

Now-a-days, there are available many commercial products known as **ANTACIDS**, which neutralize the excess of HCl secreted in the stomach. The action of antacids is based on the fact that a base can neutralize acid forming salt and water.

An essential requirement of an antacid is that it must not supply an excess of alkali which may lead to alkaline conditions, thus making the enzymes ineffective. This is achieved in the commercial antacids by incorporating other constituents which help to keep the pH within an acidic range. These drugs counteract the acid secreted in the stomach mainly to provide symptomatic relief and to a lesser extent to promote healing. Stomach keeps on emptying itself and the action of the antacid lasts only for a short while, irrespective of the dose taken. It is, therefore, important to take an antacid at frequent intervals. Commonly used antacids are Gelusil, Milk of Magnesia, Digene.

OBJECTIVE OF PROJECT

This project aims at analysing some of the commercial antacids to determine how much hydrochloric acid (of a given normality) they can neutralize.

EXPERIMENT 1

AIM: ***To analyse the given samples of commercial antacids by determining the amount of hydrochloric acid they can neutralize.***

REQUIREMENTS

Burette, pipette, titration flask, measuring flask, beakers, weight box, fractional weights, sodium hydroxide, sodium carbonate, hydrochloric acid, phenolphthalein.

PROCEDURE

1. Prepare 1 litre of approximately 0.1 N HCl solution by diluting 10 ml of the concentrated acid to one litre.
2. Similarly, make 1 litre of approximately 0.1 N NaOH solution by dissolving 4.0 g of NaOH to prepare one litre of solution.
3. Prepare 0.1 N Na_2CO_3 solution by weighing exactly 1.325 g of anhydrous sodium carbonate and then dissolving it in water to prepare exactly 250 ml of solution.
4. Standardise the HCl solution by titrating it against the standard Na_2CO_3 solution using methyl orange as indicator.
5. Similarly, standardise NaOH solution by titrating it against standardised HCl solution using phenolphthalein as indicator.
6. Powder the various samples of antacid tablets and weigh 1.0 g of each.
7. Add a specific volume of standardised HCl to each of the weighed samples taken in conical flasks. The acid should be in slight excess, so that it can neutralise all the alkaline component of the tablet.
8. Add 2 drops of phenolphthalein and warm the flask till most of powder dissolves. Filter off the insoluble material.
9. Titrate this solution against the standardised NaOH solution, till a permanent pinkish tinge is obtained. Repeat this experiment with different antacids.

OBSERVATIONS AND CALCULATIONS

Standardisation of HCl solution

Volume of 0.1 N Na_2CO_3 solution taken = 20.0 ml

S. No.	Burette readings		Volume of acid used (ml)
	Initial	Final	
1.			
2.			
3.			

Concordant reading = x ml (say)

Applying normality equation,

$$\underset{\text{(acid)}}{N_1 V_1} = \underset{\text{(base)}}{N_2 V_2}$$

$$N_1 \times x = \frac{1}{10} \times 20$$

Normality of HCl, $N_1 = \frac{2}{x}$.

Standardisation of NaOH solution.

Volume of the given NaOH solution taken = 20.0 ml.

S. No.	Burette readings		Volume of acid used (ml)
	Initial	Final	
1.			
2.			
3.			

Concordant reading = y ml (say)

Applying normality equation,

$$\underset{\text{(acid)}}{N_1' V_1'} = \underset{\text{(base)}}{N_2' V_2'}$$

$$\frac{2}{x} \times y = N_2' \times 20$$

Normality of NaOH, $N_2' = \frac{y}{10x}$.

Analysis of antacid tablets:

Weight of the antacid tablet powder = 1.0 g

Volume of HCl solution added = ... ml (say 40 ml).

Antacid	Volume of NaOH solution used for neutralising unused HCl	Volume of HCl solution used for neutralising 1.0 g of antacid matter
1. Gelusil	...... ml	...... ml
2. Milk of magnesia	...... ml	...... ml
3. Digene	...... ml	...... ml
4.	...... ml	...... ml

CONCLUSIONS

The antacid for which maximum volume of HCl is used for neutralising, is the most effective.

INVESTIGATORY PROJECT 16

Study of Constituents of an Alloy

INTRODUCTION

An **alloy** *is a homogeneous mixture of two or more metals or a metal and a non-metal.* Alloys are usually harder than their components but very often less ductile and less malleable. Thus the hardness of gold is increased by addition of copper to it. The melting point of an alloy is always lower than the melting points of the constituent metals. Other properties such as reactivity towards atmospheric oxygen and moisture, mechanical strength, ductility, colour, etc. also undergo a change when an alloy is made from its constituent metals. This change in the properties is very useful and makes alloys beneficial. Some of the alloys alongwith their composition is given below:

Alloy	*Composition*	
Brass	Copper	50–90%
	Zinc	20–40%
	+small amounts of tin, lead and iron	
Bronze	Copper	60–90%
	Tin	5–35%
	+small amounts of lead, iron and zinc	
Gun metal	Copper	85–90%
	Zinc	1–3%
	Tin	8–12%

Alloys are generally made to serve the following purposes:

(i) ***To modify chemical reactivity.*** When sodium is used as reducing agent it is too reactive to be used but its alloy with mercury, called **sodium amalgam** can be safely used as reducing agent.

(ii) ***To increase hardness.*** Hardness of gold is increased by adding copper to it.

(iii) ***To increase tensile strength.*** Nickeloy, an alloy of nickel (1%), copper (4%) and aluminium (95%) has high tensile strength.

(iv) ***To lower the melting point.*** Solder metal which is an alloy of tin (30%) and lead (70%) has very less melting point as compared to melting points of tin and lead.

(v) ***To modify the colour.*** Aluminium bronze, an alloy of copper and aluminium has beautiful golden colour.

(vi) ***To resist corrosion.*** Iron gets rusted and corroded. Its corrosion takes place with time but stainless steel which is an alloy of iron, nickel, chromium and carbon does not get rusted.

Alloys are prepared from the metals by the techniques of fusion, compression or simultaneous electro-deposition. The complete analysis of an alloy involves two steps, qualitative and quantitative analysis. In qualitative analysis, the components of the alloy are found out and in quantitative analysis their percentage composition is determined. In the present project we will carry out qualitative analysis only.

OBJECTIVE OF PROJECT

In this project, our aim is to know the various metals present in the given sample of alloy.

EXPERIMENT 1

AIM: ***To analyse a sample of brass qualitatively.***

REQUIREMENTS

China dish, test tube, funnel, filter paper and common laboratory reagents.

THEORY

A small piece of brass is dissolved in 50% nitric acid when metals get converted to their nitrates. After the removal of excess nitric acid, the solution is tested for Cu^{2+} and Zn^{+2} ions.

$$4Zn + 10HNO_3 \longrightarrow 4Zn(NO_3)_2 + N_2O\uparrow + 5H_2O$$

$$3Cu + 8HNO_3 \longrightarrow 3Cu(NO_3)_2 + 4H_2O + 2NO\uparrow$$

PROCEDURE

1. Place a small piece of brass in a china dish and heat this with minimum quantity of 50% HNO_3 so as to dissolve the piece completely.
2. Continue heating the solution till a dry solid residue is obtained.
3. Dissolve solid residue in dil. HCl and filter. Add distilled water to the filtrate.
4. Pass H_2S gas through the filtrate. A black precipitate of copper sulphide is obtained. Separate the black ppt. and keep the filtrate for the test of Zn^{+2} ions. Dissolve black ppt. by heating them with 50% HNO_3. To this solution add ammonium hydroxide solution **Appearance of deep blue colour in solution shows the presence of copper ions** in the solution.
5. To test for Zn^{+2} ions, boil the filtrate to remove H_2S gas, then add solid NH_4Cl to this and heat to dissolve NH_4Cl. Add excess of NH_4OH so that a solution is ammoniacal. Now pass H_2S gas through this ammoniacal solution. Dirty white or grey precipitate indicate zinc. Separate the precipitates and dissolve it in minimum amount of dil HCl. Boil to expel H_2S gas and add potassium ferrocyanide solution, **white or bluish white ppt. confirm Zn^{+2}** ions in the solution.

RESULT

Brass contains copper and zinc metal in it.

APPENDICES

APPENDIX 1

ATOMIC MASSES OF SOME COMMON ELEMENTS

Element	Symbol	Atomic mass
Aluminium	Al	27
Barium	Ba	137.3
Bromine	Br	79.9
Calcium	Ca	40
Carbon	C	12
Chlorine	Cl	35.5
Cobalt	Co	58.9
Chromium	Cr	52
Copper	Cu	63.6
Fluorine	F	19.0
Hydrogen	H	1
Iodine	I	127
Iron	Fe	55.8
Lead	Pb	207
Magnesium	Mg	24
Manganese	Mn	55
Mercury	Hg	201
Nickel	Ni	58.7
Nitrogen	N	14
Oxygen	O	16
Phosphorus	P	31
Potassium	K	39
Silver	Ag	108
Sodium	Na	23
Sulphur	S	32
Tin	Sn	118.7
Zinc	Zn	65

APPENDIX

2

PREPARATION OF COMMON REAGENTS USED IN THE CHEMICAL LABORATORY

APPENDIX-2A. Strengths of Concentrated Acid Solutions

Acid	*Specific Gravity*	*Percentage by Mass*	*Approximate Strength*
1. Glacial acetic acid (CH_3COOH)	1.05	99.5	17 M
2. Hydrochloric acid (HCl)	1.19	37	12 M
3. Nitric acid (HNO_3)	1.41	70	16 M
4. Sulphuric acid (H_2SO_4)	1.84	96	18 M
5. Phosphoric acid (H_3PO_4)	1.69	85	15 M

Caution: *When diluting reagents, add more concentrated reagent to the more dilute reagent* (or solvent). ***Never add water to a concentrated acid.***

APPENDIX-2B. Preparation of Dilute Solutions of Acids

Acid	*Preparation*	*Approximate Strength*
1. Acetic acid (Dil. CH_3COOH)	Dilute 285 ml of conc. CH_3COOH to 1 litre with distilled water.	5 M
2. Hydrochloric acid (Dil. HCl)	Dilute 430 ml of conc. HCl to 1 litre with distilled water.	5 M
3. Nitric acid (Dil. HNO_3)	Dilute 310 ml of conc. HNO_3 to 1 litre with distilled water.	5 M
4. Sulphuric acid (Dil. H_2SO_4)	Pour 140 ml of conc. H_2SO_4 slowly and with constant stirring into 500 ml of distilled water and dilute to 1 litre.	2.5 M

APPENDIX-2C. Preparation of Solutions of Some Bases

Acid	*Preparation*	*Approximate Strength*
1. Ammonia solution (Conc. NH_4OH)	Sp. gr. 0.88 ; 28% NH_3	15 M
2. Ammonia solution, (Dil. NH_4OH)	Dilute 335 ml of conc. NH_4OH to 1 litre with distilled water.	5 M
3. Lime water, $Ca(OH)_2$	Shake lime with distilled water and allow to stand for some time ; filter off the liquid. The clear solution is lime water.	0.02 M
4. Sodium hydroxide solution, (NaOH)	Dissolve 220 g of caustic soda in distilled water and dilute to 1 litre.	5 M

APPENDIX-2D. Preparation of Solutions of Some Salts

Reagents	Preparation	Approximate Strength
1. Ammonium acetate, CH_3COONH_4	Dissolve 231 gms of the salt in the distilled water and make the volume one litre.	3 M
2. Ammonium carbonate, $(NH_4)_2CO_3$	Dissolve 160 gms of the salt in a mixture of 140 ml of conc. ammonia solution and 860 ml of distilled water.	2 M
3. Ammonium chloride, NH_4Cl	Dissolve 270 gms of the salt in distilled water and make the volume one litre.	5 M
4. Ammonium oxalate, $(NH_4)_2C_2O_4$	Dissolve 35 gms of the salt in distilled water and make the volume one litre.	0.25 M
5. Ammonium sulphate, $(NH_4)_2SO_4$	Dissolve 132 gms of the salt in distilled water and make the volume one litre.	1 M
6. Ammonium molybdate, $(NH_4)_6Mo_7O_{24}.4H_2O$	Dissolve 45 gms of the salt in a mixture of 40 ml of conc. ammonia solution and 60 ml of water, and 120 gms of amm. nitrate and dilute with water to prepare one litre of solution.	
7. Ammonium thiocyanate, NH_4CNS	Dissolve 38 gms of the salt in water to get one litre of solution.	0.5 M
8. Yellow ammonium sulphate solution, $(NH_4)_2S_x$	Saturate 150 ml of concentrated ammonia solution with H_2S keeping the solution cold; add 1 g of flowers of sulphur and 250 ml of conc. ammonia solution, shake until the sulphur has dissolved and dilute to 1 litre.	6 M
9. Barium chloride, $BaCl_2.2H_2O$	Dissolve 61 gms of the salt in distilled water to prepare one litre of solution.	0.25 M
10. Calcium chloride, $CaCl_2.6H_2O$	Dissolve 55 gms of the hydrated salt in distilled water and make the volume of solution one litre by diluting with water.	0.25 M
11. Potassium ferrocyanide, $K_4[Fe(CN)_6]$	Dissolve 42.4 gms of the salt in water and make the volume of solution one litre by diluting with water.	0.1 M
12. Potassium iodide, KI	Dissolve 83 gms of the salt in water and make the volume of solution one litre by diluting with water.	0.1 M
13. Potassium permanganate, $KMnO_4$	Dissolve 3.2 gms of the salt in water and make the volume of solution one litre by diluting with water.	0.02 M
14. Potassium thiocyanate, KCNS	Dissolve 49 gms of the salt in water and make the volume of solution one litre by diluting with water.	0.5 M
15. Cobalt nitrate, $Co(NO_3)_2.6H_2O$	Dissolve 44 gms of the salt in water and make the volume of solution one litre by diluting with water.	0.15 M

Reagents	*Preparation*	*Approximate Strength*
16. Copper sulphate, $CuSO_4 . 5H_2O$	Dissolve 125 gms of the salt in water containing 3 ml of conc. H_2SO_4 and make the volume of solution one litre by diluting with water.	0.5 M
17. Ferric chloride, $FeCl_3 . 6H_2O$	Dissolve 55 gms of hydrated salt in distilled water containing 20 ml of conc. HCl and make the volume of solution one litre by diluting with water.	0.2 M
18. Lead acetate, $(CH_3COO)_2Pb . 3H_2O$	Dissolve 95 gms of the salt in water containing 15 ml of acetic acid and dilute to one litre.	0.25 M
19. Mercuric chloride, $HgCl_2$	Dissolve 27 gms of the salt in 1 litre of distilled water.	0.1 M
20. Silver nitrate, $AgNO_3$	Dissolve 17.0 gms of silver nitrate in distilled water and dilute to 1 litre.	0.1 M
21. Sodium acetate, $CH_3COONa . 3H_2O$	Dissolve 408 gms of the salt in distilled water and make the volume of solution one litre by diluting with water.	3 M
22. Disodium hydrogen phoshate, $Na_2HPO_4 . 12H_2O$	Dissolve 71.6 gms of the salt in water and make the volume of solution one litre by diluting with water.	0.2 M
23. Stannous chloride, $SnCl_2 . 2H_2O$	Dissolve 56 gms of the salt in 100 ml of conc. HCl and dilute to 1 litre. Keep a few pieces of tin in the bottle to prevent oxidation.	0.25 M
24. Bromine water, Br_2	Prepare saturated solution by shaking 11 ml of liquid bromine with water. Add more bromine, if necessary, to ensure a slight excess.	0.25 M
25. Chlorine water, Cl_2	Prepare saturated solution by passing chlorine gas in distilled water.	1.5 M
26. Iodine solution, I_2	Dissolve 12.7 gms of iodine in a solution of 20 gms of pure KI in 30 ml of water and dilute to 1 litre with water.	0.05 M
27. Dimethyl glyoxime reagent, $\begin{pmatrix} CH_3C = NOH \\ \vert \\ CH_3C = NOH \end{pmatrix}$	Dissolve 1 g of the substance in 100 ml of distilled rectified spirit.	
28. Sodium cobaltinitrite, $Na_3[Co(NO_2)_6]$	Dissolve 17 gms of the salt in 250 ml of water.	
29. Sodium nitroprusside, $Na_2[Fe(CN)_5NO]$	Dissolve a crystal in 5 ml of water.	
30. Potassium chromate (K_2CrO_4)	Dissolve 49 g of potassium chromate in distilled water and dilute to 1 litre.	0.25 M
31. Potassium dichromate ($K_2Cr_2O_7$)	Dissolve 147 g of potassium dichromate in distilled water and dilute to 1 litre.	0.5 M

Reagents	Preparation	Approximate Strength
32. Nessler's reagent, Alkaline solution of $K_2[HgI_4]$	Dissolved 10 gms of KI in 10 ml of ammonia free water. Add saturated $HgCl_2$ solution (60 gms per litre in small quantities at a time, with shaking until a slight permanent precipitate forms. Then add 80 ml of 9 N NaOH solution and dilute to 200 ml. Allow to stand overnight and decant off the clear liquid.	
33. Magneson	Dissolve 0.001 gm of magneson in 100 ml of N sodium hydroxide solution.	

APPENDIX-2E. Preparation of Solutions of Indicators

Reagents	Preparation	Approximate Strength
1. Phenolphthalein	Dissolve 1 g of phenolphthalein in 100 ml of alcohol.	1%
2. Methyl orange	Dissolve 1 g of methyl orange in one litre water.	0.1%
3. Methyl red	Dissolve 1 g of the solid in one litre of hot water or dissolve in 600 ml of alcohol and dilute with 400 ml of water.	0.1%
4. Starch solution	Make a paste of 1 g of starch with cold water. Pour this drop by drop into about 100 ml of boiling water and continue to boil for few minutes. Allow to stand unit cool. Decant off clear solution. If 100 g of salicylic acid is added to 100 ml of the starch solution then it keeps well for a long time.	1%
5. Litmus solution (blue)	Dissolve 10 gms of the solid in a litre of water.	1%

LOGARITHMS

	0	1	2	3	4	5	6	7	8	9	1	2	3	4	5	6	7	8	9
10	0000	0043	0086	0128	0170						5	9	13	17	21	26	30	34	38
						0212	0253	0294	0334	0374	4	8	12	16	20	24	28	32	36
11	0414	0453	0492	0531	0569						4	8	12	16	20	23	27	31	35
						0607	0645	0682	0719	0755	4	7	11	15	18	22	26	29	33
12	0792	0828	0864	0899	0934						3	7	11	14	18	21	25	28	32
						0969	1004	1038	1072	1106	3	7	10	14	17	20	24	27	31
13	1139	1173	1206	1239	1271						3	6	10	13	16	19	23	26	29
						1303	1335	1367	1399	1430	3	7	10	13	16	19	22	25	29
14	1461	1492	1523	1553	1584						3	6	9	12	15	19	22	25	28
						1614	1644	1673	1703	1732	3	6	9	12	14	17	20	23	26
15	1761	1790	1818	1847	1875						3	6	9	11	14	17	20	23	26
						1903	1931	1959	1987	2014	3	6	8	11	14	17	19	22	25
16	2041	2068	2095	2122	2148						3	6	8	11	14	16	19	22	24
						2175	2201	2227	2253	2279	3	5	8	10	13	16	18	21	23
17	2304	2330	2355	2380	2405						3	5	8	10	13	15	18	20	23
						2430	2455	2480	2504	2529	3	5	8	10	12	15	17	20	22
18	2553	2577	2601	2625	2648						2	5	7	9	12	14	17	19	21
						2672	2695	2718	2742	2765	2	4	7	9	11	14	16	18	21
19	2788	2810	2833	2856	2878						2	4	7	9	11	13	16	18	20
						2900	2923	2945	2967	2989	2	4	6	8	11	13	15	17	19
20	3010	3032	3054	3075	3096	3118	3139	3160	3181	3201	2	4	6	8	11	13	15	17	19
21	3222	3243	3263	3284	3304	3324	3345	3365	3385	3404	2	4	6	8	10	12	14	16	18
22	3424	3444	3464	3483	3502	3522	3541	3560	3579	3598	2	4	6	8	10	12	14	15	17
23	3617	3636	3655	3674	3692	3711	3729	3747	3766	3784	2	4	6	7	9	11	13	15	17
24	3802	3820	3838	3856	3874	3892	3909	3927	3945	3962	2	4	5	7	9	11	12	14	16
25	3979	3997	4014	4031	4048	4065	4082	4099	4116	4133	2	3	5	7	9	10	12	14	15
26	4150	4166	4183	4200	4216	4232	4249	4265	4281	4298	2	3	5	7	8	10	11	13	15
27	4314	4330	4346	4362	4378	4393	4409	4425	4440	4456	2	3	5	6	8	9	11	13	14
28	4472	4487	4502	4518	4533	4548	4564	4579	4594	4609	2	3	5	6	8	9	11	12	14
29	4624	4639	4654	4669	4683	4698	4713	4728	4742	4757	1	3	4	6	7	9	10	12	13
30	4771	4786	4800	4814	4829	4843	4857	4871	4886	4900	1	3	4	6	7	9	10	11	13
31	4914	4928	4942	4955	4969	4983	4997	5011	5024	5038	1	3	4	6	7	8	10	11	12
32	5052	5065	5079	5092	5105	5119	5132	5145	5159	5172	1	3	4	5	7	8	9	11	12
33	5185	5198	5211	5224	5237	5250	5263	5276	5289	5302	1	3	4	5	6	8	9	10	12
34	5315	5328	5340	5353	5366	5378	5391	5403	5416	5428	1	3	4	5	6	8	9	10	11
35	5441	5453	5465	5478	5490	5502	5515	5527	5539	5551	1	2	4	5	6	7	9	10	11
36	5563	5575	5587	5599	5611	5623	5635	5647	5658	5670	1	2	4	5	6	7	8	10	11
37	5682	5694	5705	5717	5729	5740	5752	5763	5775	5786	1	2	3	5	6	7	8	9	10
38	5798	5809	5821	5832	5843	5855	5866	5877	5888	5899	1	2	3	5	6	7	8	9	10
39	5911	5922	5933	5944	5955	5966	5977	5988	5999	6010	1	2	3	4	5	7	8	9	10
40	6021	6031	6042	6053	6064	6075	6085	6096	6107	6117	1	2	3	4	5	6	8	9	10
41	6128	6138	6149	6160	6170	6180	6191	6201	6212	6222	1	2	3	4	5	6	7	8	9
42	6232	6243	6253	6263	6274	6284	6294	6304	6314	6325	1	2	3	4	5	6	7	8	9
43	6335	6345	6355	6365	6375	6385	6395	6405	6415	6425	1	2	3	4	5	6	7	8	9
44	6435	6444	6454	6464	6474	6484	6493	6503	6513	6522	1	2	3	4	5	6	7	8	9
45	6532	6542	6551	6561	6571	6580	6590	6599	6609	6618	1	2	3	4	5	6	7	8	9
46	6628	6637	6646	6656	6665	6675	6684	6693	6702	6712	1	2	3	4	5	6	7	7	8
47	6721	6730	6739	6749	6758	6767	6776	6785	6794	6803	1	2	3	4	5	5	6	7	8
48	6812	6821	6830	6839	6848	6857	6866	6875	6884	6893	1	2	3	4	4	5	6	7	8
49	6902	6911	6920	6928	6937	6946	6955	6964	6972	6981	1	2	3	4	4	5	6	7	8

LOGARITHMS

	0	1	2	3	4	5	6	7	8	9	1	2	3	4	5	6	7	8	9
50	6990	6998	7007	7016	7024	7033	7042	7050	7059	7067	1	2	3	3	4	5	6	7	8
51	7076	7084	7093	7101	7110	7118	7126	7135	7143	7152	1	2	3	3	4	5	6	7	8
52	7160	7168	7177	7185	7193	7202	7210	7218	7226	7235	1	2	2	3	4	5	6	7	7
53	7243	7251	7259	7267	7275	7284	7292	7300	7308	7316	1	2	2	3	4	5	6	6	7
54	7324	7332	7340	7348	7356	7364	7372	7380	7388	7396	1	2	2	3	4	5	6	6	7
55	7404	7412	7419	7427	7435	7443	7451	7459	7466	7474	1	2	2	3	4	5	5	6	7
56	7482	7490	7497	7505	7513	7520	7528	7536	7543	7551	1	2	2	3	4	5	5	6	7
57	7559	7566	7574	7582	7589	7597	7604	7612	7619	7627	1	2	2	3	4	5	5	6	7
58	7634	7642	7649	7657	7664	7672	7679	7686	7694	7701	1	1	2	3	4	4	5	6	7
59	7709	7716	7723	7731	7738	7745	7752	7760	7767	7774	1	1	2	3	4	4	5	6	7
60	7782	7789	7796	7803	7810	7818	7825	7832	7839	7846	1	1	2	3	4	4	5	6	6
61	7853	7860	7868	7875	7882	7889	7896	7903	7910	7917	1	1	2	3	4	4	5	6	6
62	7924	7931	7938	7945	7952	7959	7966	7973	7980	7987	1	1	2	3	3	4	5	6	6
63	7993	8000	8007	8014	8021	8028	8035	8041	8048	8055	1	1	2	3	3	4	5	5	6
64	8062	8069	8075	8082	8089	8096	8102	8109	9116	8122	1	1	2	3	3	4	5	5	6
65	8129	8136	8142	8149	8156	8162	8169	8176	8182	8189	1	1	2	3	3	4	5	5	6
66	8195	8202	8209	8215	8222	8228	8235	8241	8248	8254	1	1	2	3	3	4	5	5	6
67	8261	8267	8274	8280	8287	8293	8299	8306	8312	8319	1	1	2	3	3	4	5	5	6
68	8325	8331	8338	8344	8351	8357	8363	8370	8376	8382	1	1	2	3	3	4	4	5	6
69	8388	8395	8401	8407	8414	8420	8426	8432	8439	8445	1	1	2	2	3	4	4	5	6
70	8451	8457	8463	8470	8476	8482	8488	8494	8500	8506	1	1	2	2	3	4	4	5	6
71	8513	8519	8525	8531	8537	8543	8549	8555	8561	8567	1	1	2	2	3	4	4	5	5
72	8573	8579	8585	8591	8597	8603	8609	8615	8621	8627	1	1	2	2	3	4	4	5	5
73	8633	8639	8645	8651	8657	8663	8669	8675	8681	8686	1	1	2	2	3	4	4	5	5
74	8692	8698	8704	8710	8716	8722	8727	8733	8739	8745	1	1	2	2	3	4	4	5	5
75	8751	8756	8762	8768	8774	8779	8785	8791	8797	8802	1	1	2	2	3	3	4	5	5
76	8808	8814	8820	8825	8831	8837	8842	8848	8854	8859	1	1	2	2	3	3	4	5	5
77	8865	8871	8876	8882	8887	8893	8899	8904	8910	8915	1	1	2	2	3	3	4	4	5
78	8921	8927	8932	8938	8943	8949	8954	8960	8965	8971	1	1	2	2	3	3	4	4	5
79	8976	8982	8987	8993	8998	9004	9009	9015	9020	9025	1	1	2	2	3	3	4	4	5
80	9031	9036	9042	9047	9053	9058	9063	9069	9074	9079	1	1	2	2	3	3	4	4	5
81	9085	9090	9096	9101	9106	9112	9117	9122	9128	9133	1	1	2	2	3	3	4	4	5
82	9138	9143	9149	9154	9159	9165	9170	9175	9180	9186	1	1	2	2	3	3	4	4	5
83	9191	9196	9201	9206	9212	9217	9222	9227	9232	9238	1	1	2	2	3	3	4	4	5
84	9243	9248	9253	9258	9263	9269	9274	9279	9284	9289	1	1	2	2	3	3	4	4	5
85	9294	9299	9304	9309	9315	9320	9325	9330	9335	9340	1	1	2	2	3	3	4	4	5
86	9345	9350	9355	9360	9365	9370	9375	9380	9385	9390	1	1	2	2	3	3	4	4	5
87	9395	9400	9405	9410	9415	9420	9425	9430	9435	9440	0	1	1	2	2	3	3	4	4
88	9445	9450	9455	9460	9465	9469	9474	9479	9484	9489	0	1	1	2	2	3	3	4	4
89	9494	9499	9504	9509	9513	9518	9523	9528	9533	9538	0	1	1	2	2	3	3	4	4
90	9542	9547	9552	9557	9562	9566	9571	9576	9581	9586	0	1	1	2	2	3	3	4	4
91	9590	9595	9600	9605	9609	9614	9619	9624	9628	9633	0	1	1	2	2	3	3	4	4
92	9638	9643	9647	9652	9657	9661	9666	9671	9675	9680	0	1	1	2	2	3	3	4	4
93	9685	9689	9694	9699	9703	9708	9713	9717	9722	9727	0	1	1	2	2	3	3	4	4
94	9731	9736	9741	9745	9750	9754	9759	9763	9768	9773	0	1	1	2	2	3	3	4	4
95	9777	9782	9786	9791	9795	9800	9805	9809	9814	9818	0	1	1	2	2	3	3	4	4
96	9823	9827	9832	9836	9841	9845	9850	9854	9859	9863	0	1	1	2	2	3	3	4	4
97	9868	9872	9877	9881	9886	9890	9894	9899	9903	9908	0	1	1	2	2	3	3	4	4
98	9912	9917	9921	9926	9930	9934	9939	9943	9948	9952	0	1	1	2	2	3	3	4	4
99	9956	9961	9965	9969	9974	9978	9983	9987	9991	9996	0	1	1	2	2	3	3	3	4

ANTILOGARITHMS

	0	1	2	3	4	5	6	7	8	9	1	2	3	4	5	6	7	8	9
.00	1000	1002	1005	1007	1009	1012	1014	1016	1019	1021	0	0	1	1	1	1	2	2	2
.01	1023	1026	1028	1030	1033	1035	1038	1040	1042	1045	0	0	1	1	1	1	2	2	2
.02	1047	1050	1052	1054	1057	1059	1062	1064	1067	1069	0	0	1	1	1	1	2	2	2
.03	1072	1074	1076	1079	1081	1084	1086	1089	1091	1094	0	0	1	1	1	1	2	2	2
.04	1096	1099	1102	1104	1107	1109	1112	1114	1117	1119	0	1	1	1	1	2	2	2	2
.05	1122	1125	1127	1130	1132	1135	1138	1140	1143	1146	0	1	1	1	1	2	2	2	2
.06	1148	1151	1153	1156	1159	1161	1164	1167	1169	1172	0	1	1	1	1	2	2	2	2
.07	1175	1178	1180	1183	1186	1189	1191	1194	1197	1199	0	1	1	1	1	2	2	2	2
.08	1202	1205	1208	1211	1213	1216	1219	1222	1225	1227	0	1	1	1	1	2	2	2	3
.09	1230	1233	1236	1239	1242	1245	1247	1250	1253	1256	0	1	1	1	1	2	2	2	3
.10	1259	1262	1265	1268	1271	1274	1276	1279	1282	1285	0	1	1	1	1	2	2	2	3
.11	1288	1291	1294	1297	1300	1303	1306	1309	1312	1315	0	1	1	1	2	2	2	2	3
.12	1318	1321	1324	1327	1330	1334	1337	1340	1343	1346	0	1	1	1	2	2	2	2	3
.13	1349	1352	1355	1358	1361	1365	1368	1371	1374	1377	0	1	1	1	2	2	2	3	3
.14	1380	1384	1387	1390	1393	1396	1400	1403	1406	1409	0	1	1	1	2	2	2	3	3
.15	1413	1416	1419	1422	1426	1429	1432	1435	1439	1442	0	1	1	1	2	2	2	3	3
.16	1445	1449	1452	1455	1459	1462	1466	1469	1472	1476	0	1	1	1	2	2	2	3	3
.17	1479	1483	1486	1489	1493	1496	1500	1503	1507	1510	0	1	1	1	2	2	2	3	3
.18	1514	1517	1521	1524	1528	1531	1535	1538	1542	1545	0	1	1	1	2	2	2	3	3
.19	1549	1552	1556	1560	1563	1567	1570	1574	1578	1581	0	1	1	1	2	2	3	3	3
.20	1585	1589	1592	1596	1600	1603	1607	1611	1614	1618	0	1	1	1	2	2	3	3	3
.21	1622	1626	1629	1633	1637	1641	1644	1648	1652	1656	0	1	1	2	2	2	3	3	3
.22	1660	1663	1667	1671	1675	1679	1683	1687	1690	1694	0	1	1	2	2	2	3	3	3
.23	1698	1702	1706	1710	1714	1718	1722	1726	1730	1734	0	1	1	2	2	2	3	3	4
.24	1738	1742	1746	1750	1754	1758	1762	1766	1770	1774	0	1	1	2	2	2	3	3	4
.25	1778	1782	1786	1791	1795	1799	1803	1807	1811	1816	0	1	1	2	2	2	3	3	4
.26	1820	1824	1828	1832	1837	1841	1845	1849	1854	1858	0	1	1	2	2	3	3	3	4
.27	1862	1866	1871	1875	1879	1884	1888	1892	1897	1901	0	1	1	2	2	3	3	3	4
.28	1905	1910	1914	1919	1923	1928	1932	1936	1941	1945	0	1	1	2	2	3	3	4	4
.29	1950	1954	1959	1963	1968	1972	1977	1982	1986	1991	0	1	1	2	2	3	3	4	4
.30	1995	2000	2004	2009	2014	2018	2023	2028	2032	2037	0	1	1	2	2	3	3	4	4
.31	2042	2046	2051	2056	2061	2065	2070	2075	2080	2084	0	1	1	2	2	3	3	4	4
.32	2089	2094	2099	2104	2109	2113	2118	2123	2128	2133	0	1	1	2	2	3	3	4	4
.33	2138	2143	2148	2153	2158	2163	2168	2173	2178	2183	0	1	1	2	2	3	3	4	4
.34	2188	2193	2198	2203	2208	2213	2218	2223	2228	2234	1	1	2	2	3	3	4	4	5
.35	2239	2244	2249	2254	2259	2265	2270	2275	2280	2286	1	1	2	2	3	3	4	4	5
.36	2291	2296	2301	2307	2312	2317	2323	2328	2333	2339	1	1	2	2	3	3	4	4	5
.37	2344	2350	2355	2360	2366	2371	2377	2382	2388	2393	1	1	2	2	3	3	4	4	5
.38	2399	2404	2410	2415	2421	2427	2432	2438	2443	2449	1	1	2	2	3	3	4	4	5
.39	2455	2460	2466	2472	2477	2483	2489	2495	2500	2506	1	1	2	2	3	3	4	5	5
.40	2512	2518	2523	2529	2535	2541	2547	2553	2559	2564	1	1	2	2	3	4	4	5	5
.41	2570	2576	2582	2588	2594	2600	2606	2612	2618	2624	1	1	2	2	3	4	4	5	5
.42	2630	2636	2642	2649	2655	2661	2667	2673	2679	2685	1	1	2	2	3	4	4	5	6
.43	2692	2698	2704	2710	2716	2723	2729	2735	2742	2748	1	1	2	3	3	4	4	5	6
.44	2754	2761	2767	2773	2780	2786	2793	2799	2805	2812	1	1	2	3	3	4	4	5	6
.45	2818	2825	2831	2838	2844	2851	2858	2864	2871	2877	1	1	2	3	3	4	5	5	6
.46	2884	2891	2897	2904	2911	2917	2924	2931	2938	2944	1	1	2	3	3	4	5	5	6
.47	2951	2958	2965	2972	2979	2985	2992	2999	3006	3013	1	1	2	3	3	4	5	5	6
.48	3020	3027	3034	3041	3048	3055	3062	3069	3076	3083	1	1	2	3	4	4	5	6	6
.49	3090	3097	3105	3112	3119	3126	3133	3141	3148	3155	1	1	2	3	4	4	5	6	6

ANTILOGARITHMS

	0	1	2	3	4	5	6	7	8	9	1	2	3	4	5	6	7	8	9
.50	3162	3170	3177	3184	3192	3199	3206	3214	3221	3228	1	1	2	3	4	4	5	6	7
.51	3236	3243	3251	3258	3266	3273	3281	3289	3296	3304	1	2	2	3	4	5	5	6	7
.52	3311	3319	3327	3334	3342	3350	3357	3365	3373	3381	1	2	2	3	4	5	5	6	7
.53	3388	3396	3404	3412	3420	3428	3436	3443	3451	3459	1	2	2	3	4	5	6	6	7
.54	3467	3475	3483	3491	3499	3508	3516	3524	3532	3540	1	2	2	3	4	5	6	6	7
.55	3548	3556	3565	3573	3581	3589	3597	3606	3614	3622	1	2	2	3	4	5	6	7	7
.56	3631	3639	3648	3656	3664	3673	3681	3690	3698	3707	1	2	3	3	4	5	6	7	8
.57	3715	3724	3733	3741	3750	3758	3767	3776	3784	3793	1	2	3	3	4	5	6	7	8
.58	3802	3811	3819	3828	3837	3846	3855	3864	3873	3882	1	2	3	4	4	5	6	7	8
.59	3890	3899	3908	3917	3926	3936	3945	3954	3963	3972	1	2	3	4	5	5	6	7	8
.60	3981	3990	3999	4009	4018	4027	4036	4046	4055	4064	1	2	3	4	5	6	6	7	8
.61	4074	4083	4093	4102	4111	4121	4130	4140	4150	4159	1	2	3	4	5	6	7	8	9
.62	4169	4178	4188	4198	4207	4217	4227	4236	4246	4256	1	2	3	4	5	6	7	8	9
.63	4266	4276	4285	4295	4305	4315	4325	4335	4345	4355	1	2	3	4	5	6	7	8	9
.64	4365	4375	4385	4395	4406	4416	4426	4436	4446	4457	1	2	3	4	5	6	7	8	9
.65	4467	4477	4487	4498	4508	4519	4529	4539	4550	4560	1	2	3	4	5	6	7	8	9
.66	4571	4581	4592	4603	4613	4624	4634	4645	4656	4667	1	2	3	4	5	6	7	9	10
.67	4677	4688	4699	4710	4721	4732	4742	4753	4764	4775	1	2	3	4	5	7	8	9	10
.68	4786	4797	4808	4819	4831	4842	4853	4864	4875	4887	1	2	3	4	6	7	8	9	10
.69	4898	4909	4920	4932	4943	4955	4966	4977	4989	5000	1	2	3	5	6	7	8	9	10
.70	5012	5023	5035	5047	5058	5070	5082	5093	5105	5117	1	2	4	5	6	7	8	9	11
.71	5129	5140	5152	5164	5176	5188	5200	5212	5224	5236	1	2	4	5	6	7	8	10	11
.72	5248	5260	5272	5284	5297	5309	5321	5333	5346	5358	1	2	4	5	6	7	9	10	11
.73	5370	5383	5395	5408	5420	5433	5445	5458	5470	5483	1	3	4	5	6	8	9	10	11
.74	5495	5508	5521	5534	5546	5559	5572	5585	5598	5610	1	3	4	5	6	8	9	10	12
.75	5623	5636	5649	5662	5675	5689	5702	5715	5728	5741	1	3	4	5	7	8	9	10	12
.76	5754	5768	5781	5794	5808	5821	5834	5848	5861	5875	1	3	4	5	7	8	9	11	12
.77	5888	5902	5916	5929	5943	5957	5970	5984	5998	6012	1	3	4	5	7	8	10	11	12
.78	6026	6039	6053	6067	6081	6095	6109	6124	6138	6152	1	3	4	6	7	8	10	11	13
.79	6166	6180	6194	6209	6223	6237	6252	6266	6281	6295	1	3	4	6	7	9	10	11	13
.80	6310	6324	6339	6353	6368	6383	6397	6412	6427	6442	1	3	4	6	7	9	10	12	13
.81	6457	6471	6486	6501	6516	6531	6546	6561	6577	6592	2	3	5	6	8	9	11	12	14
.82	6607	6622	6637	6653	6668	6683	6699	6714	6730	6745	2	3	5	6	8	9	11	12	14
.83	6761	6776	6792	6808	6823	6839	6855	6871	6887	6902	2	3	5	6	8	9	11	13	14
.84	6918	6934	6950	6966	6982	6998	7015	7031	7047	7063	2	3	5	6	8	10	11	13	15
.85	7079	7096	7112	7129	7145	7161	7178	7194	7211	7228	2	3	5	7	8	10	12	13	15
.86	7244	7261	7278	7295	7311	7328	7345	7362	7379	7396	2	3	5	7	8	10	12	13	15
.87	7413	7430	7447	7464	7482	7499	7516	7534	7551	7568	2	3	5	7	9	10	12	14	16
.88	7586	7603	7621	7638	7656	7674	7691	7709	7727	7745	2	4	5	7	9	11	12	14	16
.89	7762	7780	7798	7816	7834	7852	7870	7889	7907	7925	2	4	5	7	9	11	13	14	16
.90	7943	7962	7980	7998	8017	8035	8054	8072	8091	8110	2	4	6	7	9	11	13	15	17
.91	8128	8147	8166	8185	8204	8222	8241	8260	8279	8299	2	4	6	8	9	11	13	15	17
.92	8318	8337	8356	8375	8395	8414	8433	8453	8472	8492	2	4	6	8	10	12	14	15	17
.93	8511	8531	8551	8570	8590	8610	8630	8650	8670	8690	2	4	6	8	10	12	14	16	18
.94	8710	8730	8750	8770	8790	8810	8831	8851	8872	8892	2	4	6	8	10	12	14	16	18
.95	8913	8933	8954	8974	8995	9016	9036	9057	9078	9099	2	4	6	8	10	12	15	17	19
.96	9120	9141	9162	9183	9204	9226	9247	9268	9290	9311	2	4	6	8	11	13	15	17	20
.97	9333	9354	9376	9397	9419	9441	9462	9484	9506	9528	2	4	7	9	11	13	15	17	20
.98	9550	9572	9594	9616	9638	9661	9683	9705	9727	9750	2	4	7	9	11	13	16	18	20
.99	9772	9795	9817	9840	9863	9886	9908	9931	9954	9977	2	5	7	9	11	14	16	18	20